More Praise for *Song of Silence*

"Cynthia Ruchti's enthralling novels are always on my automatic buy list. Her beautifully written stories take me on a journey through my own heart and soul. Highly recommended!" —**Colleen Coble**, author of the Sunset Cove series and the Rock Harbor novels

"Sometimes God shatters a dream because what he will create out of the broken pieces will be far more beautiful and useful than the original. *Song of Silence* is the glorious telling of one family's broken dreams transformed by hope. Once again, Cynthia Ruchti delights us with poetic words woven into a faith-building, heart-tugging story." —**Becky Melby**, The Lost Sanctuary Series

"The story within these pages mirrors my own on many levels. From music education to special needs; from heartache to the reality of relationships, Ruchti nails every note with crafted precision. From downbeat to finale, this book rang deeply in my heart." —**Richard Pickrell**, music education specialist, parent of a special needs child

"Have you ever experienced the loss of a dream…followed by the loss of that very thing (talent, ability, character trait) within yourself that made the dream possible? Cynthia Ruchti chronicles that loss in the life of Lucy. She deftly depicts the feelings of unreality, depression, and the uncanny ability of life to follow up one knockout punch with a string of other shattering blows. Yet as despair hovers menacingly close, Cynthia injects her trademark perspective of a life 'hemmed in hope.'" —**Kathy Kexel**, on-air host, 88.5 The Family Radio

"Author Cynthia Ruchti once again tackles a life problem and delivers a hope-filled outcome while being honest about the struggle. In *Song of Silence*, readers will ask themselves, 'What would I do if the thing that makes me feel like I matter in life was taken away and I had to start over? Where would I find my significance?' Cynthia's characters (of all ages) deliver real-to-life dialogue and dilemmas. The scenes balance funny, fear, and futility, resulting in renewed faith." —**Kathy Carlton Willis**, pastor's wife, women's ministry director, author of the books *Grin with Grace* and *Speaker to Speaker*

"Cynthia Ruchti has created vivid characters who react to change in the same ways we might respond. Put *Song of Silence* on your must-read list." —**Karen Porter**, author, international speaker, coach, successful businesswoman

"A beautifully written story that hits close to home in many ways. When life isn't going the way you'd planned—when the melody seems to fade away mid-song—Ruchti's book shares a message of hope and healing that resonates." —**Jenness Walker**, author of *Double Take*

"A warm reminder that God works unexpectedly in the silence of our lives. As a teacher who has experienced reduction in force (RIF), the emotions of the main character, Lucy, proved transparent and real. This book gripped my heart from the first page to the last with the reality of the situation. It taught me to 'keep playing in the silence.'" —**Cleo Lampos**, M.Ed., special education teacher, speaker, author of Teachers of Diamond Project School Series and *Teaching Diamonds in the Tough*.

"With her exquisitely engaging style, each word in perfect pitch, Cynthia Ruchti creates a beautiful story of healing and humor and hope. A story for the riffed teacher, for the frustrated wife, for the mother of adult children. A story for everyone whose life failed to proceed according to plan. Vivid characters, wry wit, and gentle truth held me captive from the first page to the last. This is a book that will linger in my heart for a long time!" —**Carrie Schmidt**, wife, English-as-a-second-language teacher, book reviewer, participant in a life not proceeding according to plan

"Love, faith, and understanding within a family is at the heart of *Song of Silence*. Cynthia asked to use Lucy and Charlie as names for the main characters because I (Lucy) was Cynthia's elementary music teacher. She was a 'sparkplug' and gave energy to all aspects of my teaching. I could relate to the characters' personality conflicts in several scenes in *Song of Silence* and to their determination to resolve them positively, despite the challenges and disappointments that accompany life's song." —**Lucy Ramshaw**, retired music teacher

"Cynthia has captured the heart of the main character's struggle to redirect her focus through more challenging events than she could ever have dreamed. *Song of Silence* spoke to me personally as I seek God's direction in a new phase of my own life." —**Helen McCormack**, Wycliffe Missionary and former special education teacher

"*Song of Silence* is a journey of heartache, soul-searching, passion-seeking, courage, and healing during a 'timed rest' in life that silences one's heart song. It's written with a tenderness that leaves you greatly blessed." —**Lynae Frank Holman**, teacher of the deaf and hard of hearing

song of silence

CYNTHIA RUCHTI

Guideposts
Danbury, Connecticut

Song of Silence

Copyright © 2016 by Cynthia Ruchti

All rights reserved. No part of this work may be reproduced or transmitted in any form or by any means, except as may be expressly permitted by the U.S. Copyright Law. For Permissions contact: PermissionsEditor@guideposts.org

The persons and events portrayed in this work of fiction are the creations of the author, and any resemblance to persons living or dead is purely coincidental.

Published by Guideposts a Church Corporation
100 Reserve Road, Suite E200
Danbury, CT 06810-5212
Guideposts.org

This Guideposts edition is published by special arrangement with Abingdon Press.

All scripture quotations unless noted otherwise are taken from the Common English Bible. Copyright © 2011 by the Common English Bible. All rights reserved. Used by permission. www.CommonEnglishBible.com.

Psalm 51:10 on page 145 is taken from The Authorized (King James) Version. Rights in the Authorized Version in the United Kingdom are vested in the Crown. Reproduced by permission of the Crown's patentee, Cambridge University Press.

Printed in the United States of America
10 9 8 7 6 5 4 3 2 1

*To those who make space for music
and those who leave room for silence*

Acknowledgments

I owe a debt of gratitude to the music educators who nurtured a lifelong appreciation for the power and influence of music. Lucy's passion in this story is a reflection of a real-life Lucy, who had such an impact on my life that the songs she taught me—and the emotional connection to the songs—are tucked in my soul's archives forever. Thank you, Lucy. (And thank you for giving permission to name my characters after you and your Charlie.)

My father—whom I consider among the top music educators of all time—didn't live long enough to know I would write a story he influenced so strongly. Your legacy lives on in far more lives than just mine, Dad.

I appreciate the riffed teachers who shared their heartache with me. Each state or individual school has procedures and fallout unique to its protocol. Forgive me if your experience didn't look like what is written in these pages. Please know, though, that your story matters.

A key plot point owes its drama to a woman who walked the painful path and found at least a measure of redemption in its inclusion here. Thank you for your insights, Jennifer Zarifeh Major. I'm grateful your song returned. Thank you, too, for showing me details about some forms of silence I wouldn't have known without your family's experience.

Thank you, Sara and Richard, for sharing the gift of your son with our church community for those treasured years. He enriched these pages.

I'm sure I've said it before, but my gratitude constantly runs at flood stage for my agent, Wendy Lawton; my editor, Ramona Richards; and the Abingdon Press family. Every new project we work on together increases my appreciation for how hard you work and for the heart you put into your roles.

Becky Melby, I've told you in person, but say it publicly here, that I don't know how I would have gotten this book written without your encouragement, your critiques, your prayers, and your relentless cheerleading.

The connect group with which my husband and I identify has invested much in every book I write. This one is no exception. They pray and rally, not knowing what the story is about until they hold it in their hands and pray over it again.

My family's sense of humor and devotion to one another—and to the important things in life—show up within these pages. Reader, it's a joy to share them with you in this way.

Giver of Song, please accept my gratitude for what both song and silence have meant to me.

Only when you drink from the river of silence shall you indeed sing.

—Kahlil Gibran

1

Lucy removed her glasses and watched Ellie's thin, thirteen-year-old fingers splay against the girl's too-flat stomach.

"Try it," Lucy said.

"I don't have much breath."

"I know." The confession drilled so much deeper than it would have coming from any of Lucy's other students. "Please try."

She watched as Ellie struggled to fill her scarred lungs from the bottom without moving her upper chest or shoulders. The girl's hand moved an inch.

"Now, inhale and exhale without letting your hand move at all."

"I can't."

Lucy tilted her head, eyebrows raised, wordlessly urging a response from Ellie.

Ellie smiled. "Time to be brave? Braver than I feel?"

"Right." Lucy traced the girl's line of sight to one of the dozens of motivational posters on the wall. *Be Brave. Braver than you feel.* Next to it, *Right or wrong, blow it strong.* Beside that one, *Practice doesn't make perfect. It makes possible.* Lucy's favorite, *Just so you know, dogs don't eat music homework.*

"Deep breath from the bottom of your lungs. Push your abdomen out to allow air in. Hold it. Now two small breaths in and out without moving your hand. There! You did it!"

Ellie pressed her lips together but couldn't stop the smile that overrode her efforts. "I didn't think I could."

"Now, let's try that technique for these four measures." Lucy pointed to the sheet on the music stand. "Keep that expansion in your tummy, even though you'll have to breathe. See if it doesn't help you maintain that beautiful tone you've been working on."

The girl raised the silver flute to her pursed lips, a mix of eagerness and skepticism on her face. She exaggerated the movement of her abdomen, her striped shirt proving her obedience, and played the specified measures. Ellie's eyes flashed her reaction before she lowered her flute. "That," she said, "was awesome!"

Tears tickled Lucy's sinuses. "Yes, it was."

"Does that work with singing, too? Could I join choir next year? Is there room for me?"

Laughter poured out of Lucy's mouth, but it originated in her heart. "Four brilliant measures and you're ready to tackle singing, too?" As quickly as the laughter erupted, it died. *Her* choir? Next year?

"My doctor says he owes you." Ellie's flute lay in her lap, the thin fingers cradling it. She stifled most of a cough. "He says he never would have thought of music as cystic fibrosis therapy."

I never thought my first chair flutist would muscle through CF to keep playing. "I'm glad it's helping."

"GDBD," she said, running her fingers over the instrument.

"Good days, bad days?"

Ellie looked up. "Do you *text*?" Incredulity.

Lucy took no offense. Even at a few months shy of fifty-six, she must have seemed ancient to a thirteen-year-old. Despite her sassy haircut. And artsy earrings, thanks to Ania's jewelry-making skills. "Is today a good day, Ellie?"

The girl lifted her flute then pointed to the line of notes on the page, as a pool player might point to the pocket where she intended the eight ball to land. "Mrs. Tuttle, any day I'm breathing is considered a good day." She inhaled without moving her shoulders and played the measures as if running a victory lap. Which she would likely never do. Run.

Lucy was three hours away from another school board budget-cut meeting. Could she keep breathing? The discussion had crept too close to destroying scenes like this one with Ellie. Only Lucy's dogged sense of propriety had kept her from storming the school board's line of tables and chairs last time. If it crept much closer...

Lucy turned her attention back to her admiration for a thirteen-year-old's breathless ability to muscle through.

When Ellie's smile left the room, Lucy retreated to her cramped office at the end of the line of three small practice rooms. She stared at the screen of her laptop, open to her calendar. The school day was over, but her list of duties hadn't shrunk. Spring concert next week. She needed to sneak in another announcement for the Woodbridge radio station and create another mass text message for the parents and grandparents who paid more attention to texts than they did the school's weekly newsletter.

Charlie said he'd eat at Bernie's tonight. She could work straight through until the budget-cut meeting if she wanted. He'd meet her there. Why couldn't he be the one to speak up in a public forum? Why did he slip into it'll-all-work-out mode when her life stood in the crosshairs? So much for knight on a white horse. But he would be there. She didn't have to wonder if he'd show up.

She needed a new office chair. One that didn't groan when she moved. Or was that sound coming from her soul?

Two hours later she pushed away from her desk and closed the lid of her laptop. She shouldn't head into the meeting with an empty stomach. But it might be *emptied* by the outcome of the gathering, barring divine intervention. So she had no clear choice.

Divine intervention. Nothing short would move a woman like Evelyn Schindler, who approached budget cuts with the ruthlessness of a self-guided chain saw.

"It's difficult to take your perspective seriously," Evelyn "Chain Saw" Schindler said, leaning too far into her microphone. She jerked back, as if she'd chipped a tooth in her enthusiasm to make her point. Straightening her posture to a stiffness well past at ease, she added, "You're the music instructor, Mrs. Tuttle. Is there any doubt where you'd stand on the issue? Those overly passionate add a skewed perspective to the subject at hand. I think we can all agree on that."

The woman nodded to the board members on her left and right, some of whom nodded back. Others dropped their gaze. And their opportunity to disagree. Lucy's friends, some of them. People she'd known since her father held the position she clung to now with a free-climber's fingertips-only grip.

Nothing but air at her back. Hundreds of feet above the sun-baked canyon floor. Toes pretending the quarter-inch crack in the rock is enough. Fingertips stretching skyward, muscles straining to hold out for a dependable ledge.

"Mrs. Tuttle." The board president's voice sounded like one reserved for the detention room.

"What?"

"You can lower your hand. We've heard your opinion. There are others waiting to have their say."

Lower her—? That's what she needed. Another reason to be embarrassed. She slipped her hand down and bent to retrieve her bottled water from the floor. It bought her enough time to refocus.

Charlie patted her knee. Could have been his "Steady, girl," or "There, there now," or "Way to go, honey." Probably one of the first two options.

The next speaker's rabbit trail wandered so far afield, Lucy feared his point had already crossed into the next county without him. Hope followed—a string of community members, many of them parents of her students—voicing logical, well-expressed reasons to look someplace other than art and music for the necessary cuts.

For a small town like Willowcrest to maintain a private school without federal funding for more than four decades, they'd danced to the edge of tough decisions more than once. The seriously sports-minded usually transferred to a Woodbridge school. But thanks in large part to Lucy's father's influence, the music program *kept* students in Willowcrest.

That point worked its way into the next speaker's impassioned plea. Ellie's mom. And the next. A parent from a student long graduated.

Lucy watched as the panel of school board members scribbled notes—or graffiti—onto memo pads. Evelyn Schindler's shoulders sagged. Could the tide be turning?

The next community member given the floor presented an anti-music-education argument so flawed, it drew snickers from the crowd. He grabbed his frayed baseball cap from his folding chair, pointed toward the board and said, "You know we got no choice." His exit brought Lucy relief she assumed was shared by others, judging from the expressions on the faces of more than half of the attendees.

Who was that leaning against the wall near the exit? A reporter? Mid-twenties, she guessed. Not someone she'd seen around the community, that she could remember. From where Lucy sat, she could pay attention to the proceedings and keep an eye on the intense young man, too, if she turned a few degrees in her chair. Charlie took that gesture as a reason to put his arm around his wife.

"When did Olivia get here?" she whispered into the better of Charlie's ears.

"She's here?" He swiveled his head toward the standing room only spot not far from the reporter. He waved like a second grader might wave to his parents in the audience.

"Charlie!"

"What?"

Evelyn Schindler made her microphone squeal. "If we could have everyone's attention? Time limits being as they are, we're going to need to wrap this up for tonight. The board will agree with me, I'm sure—"

Don't they always?

"—that we've been given more than enough food for thought in this matter. As always, we remain open to your comments via e-mail or personal contact. Let's call it a night, shall we, folks?"

Well. No pronouncement of doom. Had Lucy's music program dodged another wrecking ball for the moment? She glanced back toward Olivia, who stood talking with the reporter guy. What did her daughter have to say to him? What was he asking? Lip-reading would come in handy at a time like this.

Part of Lucy's brain allowed her to converse with community members voicing their ongoing support while she watched Olivia and the note taker leave. Together.

Lucy texted Olivia on the short drive home. "Cute guy. Someone special?"

Olivia texted back, "Could be. We'll see."

"You coming over?"

"Heading back to Woodbridge now. See you soon. Praying for you, Mom."

"Thanks."

Lucy had to admit texting came in handy once in a while. It kept her better connected with her kids.

"Nice of Olivia to show up," Charlie said, adjusting his rear view mirror.

"I haven't talked to her for a couple of days. Thought maybe she'd spend the night."

"She isn't?"

Lucy unlooped the lightweight infinity scarf around her neck and tucked it into her purse. "Heading back."

"I should have asked her to go out for frozen custard with us."

"Them."

"What?"

"Should have asked them. She was with someone."

Charlie's eyebrows registered his surprise.

"You're not suggesting you want to stop for custard, are you?"

"You don't want to?" His voice wavered as if she'd told him he couldn't have a puppy.

"Could we just go home?"

"These meetings take a lot out of you, don't they, LucyMyLight?"

If you'd ever had a passion, Charlie, a job you were invested in, a career or interest that meant as much to you as mine does to me, you'd get it. You'd understand why nights like this are reason enough for a heart attack. "I'm tired. And I still have work to do on next week's concert schedule."

"Can we go through the drive-thru? I had my heart set on—"

"Sure." One day, she'd stop saying *sure* when she meant *no*.

It was probably too much to expect the school board members to attend the students' spring concert. Boycotted it, apparently. Lucy didn't mind that Evelyn Schindler stayed away last night. She rarely showed. But some of the missing board members were parents or grandparents of students in Lucy's band and chorus. Last concert of the year. They couldn't all be under the weather.

Community support made up for their absence. Who does a standing ovation for a K-8 concert? Too bad the members of the school board hadn't seen it.

It wouldn't be wrong if Lucy sent a copy of the video to each of their in-boxes, would it? It would be informational, inclusive, and *thoughtful* of her.

With the new school day an hour away from starting, she let herself into the still quiet music room, settled into her office, opened her laptop and calendar, and made a note to send the file when the tech team had it ready. The afterglow of the concert lingered. She'd heard every note in her mind through the night, seen the faces of the young people lit from within as the music took hold in their souls. *And that—budget-fussy people—is why you can never cut this program.*

Her computer dinged. E-mail. From Ania.

Before she opened it, she keyed in another note to herself to have the music students write a group thank-you to the art students whose work lined the lobby for last night's concert. Ania might be young, but she'd made great strides with her students her first year of teaching.

Lucy clicked the e-mail.

"Did you open your mail yet?"

The letters and catalogs sat on the edge of her desk. With so much accomplished electronically, her stack of mail rarely amounted to much anymore. She thumbed through it. One envelope wasn't postmarked. Hand-addressed. A thank-you from one of the faculty members?

It had been sealed in one spot only—at the point of the V of the back flap. Who hadn't wanted to waste saliva?

Lucy read the first five words before the sound of a distant chain saw stopped her.

The two-mile drive from the Willowcrest School to her house on Cottonwood had never felt like a commute until today. Innocent clouds seemed sinister. Her body registered every groove or divot in the pavement despite the layers of automotive steel, plastic, and upholstery separating her from them. She was fourth to arrive at all three of the four-way stops. Hollowness expanded like out-of-control yeast dough the farther she drew from the school.

May usually represents hope reborn in the Upper Midwest. Winter laid to rest. Spring-almost-summer putting down taproots. Vivid colors. Lilied and peonied air. Leaves so fresh, they look damp. A vibration of exuberant life that thrums like a baby robin's heartbeat.

Despite the only partly cloudy sky, Lucy saw dull colors, faded, fogged over. She heard only muted tones. The smell of her car's citrus air freshener choked her. While stopped at another stop sign, she ripped the freshener from its resting place and jammed it into the litterbag.

Was it just her, or did the street sign on Cottonwood look tilted? Not much. Just enough to notice. And the mailbox leaned the opposite way. Dr. Seussian.

She turned off the engine and stared at the front door of her house. What made her think she could pull off a turquoise door on a moss green house? Ania's idea. Ania didn't know everything. But who was Lucy to talk?

In a motion so automatic she didn't have to think about it—which was good on a day like today—Lucy pocketed her keys, slid her purse and tote bag from the passenger seat, and exited the car in one nearly smooth motion. The glaringly bright turquoise door swung open as she reached for the knob.

"I found my passion!" Charlie's graying eyebrows danced. Nothing else moved. A statue of a man with jive eyebrows.

"Happy for you. Is it okay if I get all the way into the house before you tell me the rest of your story?" Lucy nudged her husband with her shoulder as she scooted past his Ed Asner form. How much could a doorframe swell in mid-May's premature humidity? Were the walls swollen too? The whole house felt smaller. Shrunken.

Charlie stayed on her heels as she deposited her *2014 Milner County Teacher of the Year* tote bag and leather hobo purse on the repurposed vanity/hall table. "Charlie. Some space?"

"Don't you want to know what it is?" Charlie's head tilt reminded Lucy of a terrier pup they'd seen in the neighborhood. Cute, on a puppy. Mildly cute on the sometimes-annoying love of her life.

"Can I have a minute to acclimate?" She cupped his jaw and kissed the tip of his decades-familiar nose. "Not my best day, Charlie."

"Mine," he said, pulling her close, "got decidedly better when you walked through the door."

"You read that line in a book, didn't you." Her heart warmed a degree or two in spite of the icy talons holding it in their grip.

He pulled back. "Am I that transparent?"

"Like a sixth-grader's homework excuse."

"I never claimed to be a romantic."

She tugged at the silver curling in front of his ears. "Time for a haircut, young man."

"My barber had a bad day, I hear. Not sure I trust her with scissors." Charlie pressed his palms to the sides of his head. "I can't afford a distracted stylist. Or shorter ears." His grin would have seemed impish on an ordinary day.

"You could spring for a professional barber once in a while, you know. We can"—*could*, she silently corrected—"afford it." She turned to the stack of mail on the table. Not yet. She wasn't ready to say the words. "And they're shears. We semiprofessionals don't call them scissors. They're shears."

"You bought them at Walmart."

"Touché."

The fencing foil—lodged in her throat since eight hours earlier—slipped farther down. To the hilt.

She'd have to tell him.

So... after all these years, he'd found his life's passion. On the day she lost hers.

"Worms, Lucy."

She'd only managed to kick off one shoe before he spewed his news. Hers would have to wait. "You have worms?"

"Not yet. But I will."

"You need to stay away from the pet rescue center for a while." *Charlie, Charlie, Charlie.* "And keep hand sanitizer in your truck." Second shoe. *Coffee. Need coffee.*

He bent to line her abandoned leather mules in a row with the other shoes on the mat beside the entry table. Who knew retirement would turn him into a neat freak? So not his style.

"Worm farm, LucyMyLight."

The nickname he'd started using when they dated in college had never seemed aggravating before. But it felt as uncomfortable as a fiberglass sweater today. She blamed it on the barbed letter.

He took her hand as he had so often over the years and tugged her toward the kitchen. She slumped into the chair he pulled out for her, then forced her posture into a neutral, unreadable position. The man was pouring her a cup of coffee, eyebrows still dancing, and launching into a personal infomercial about worm farms. Now was not the time to collapse.

"I think this is it, Lucy. The thing I can get passionate about." He slid her treble-clef mug toward her and lowered himself into the chair opposite hers. The pale beige brew in his nondescript coffee mug looked more like anemic chocolate milk rather than the Costa Rican mahogany that filled hers.

Charlie, Charlie, Charlie. "You want to raise bait? Fishing bait?"

"Now, see? That's the common misperception—that worm farms are only good for producing night crawlers. Which, honestly, is reason enough by itself. I think anyone would admit that."

She held the coffee under her nose. Too hot to drink. The aroma helped her mood about as much as a photograph of an antidepressant.

"A high-tech worm farm can produce—"

Had he just used *high-tech* and *worm farm* in the same sentence? His words squirmed in the air between them.

"—and soil enrichment, because of their... you know... feces."

Collapses hold to no schedules. She pushed her coffee out of the way and laid her head on the table like a toddler falling asleep in her SpaghettiOs.

"Lucy! What's wrong? Is it your heart?"

Her heart? She was not that old. And, sure, heart attacks knew no age limits. But really? His first thought was her heart? "No. Yes." Her words disappeared into the tabletop. The scent of oranges cocooned around her. He'd bought the off-brand furniture polish again.

His chair legs scraped the ceramic tile. "I'll get a baby aspirin." She heard his footsteps pounding toward the powder room. Who knew he could move so fast?

She lifted her head long enough to say, "I'm not having a heart attack."

"Stroke?"

Such a helpful man. "No. Close, but no." She propped her elbows on the table and cupped her forehead in both hands. Her skull still seemed two sizes too small.

"Look, do I call 9-1-1 or not?" Charlie's voice shifted from panic mode to irritation.

"Not. They can't do anything about this."

She knew without opening her eyes that he'd set the portable phone back in its base. A moment later, she felt his hand rubbing her upper back, tentatively, as if unsure if touching her would make it worse. "Lucy..."

The only warmth left in her lay across her shoulders, under his hand. "I lost my passion."

"For... me?"

How could he think that after all these years? She sat back and leaned her head against his Ed Asnerness. She could hear his traditional mid-afternoon popcorn digesting. "I lost my job," she said, choking on the words. "They cut the program."

He stepped away without warning. Her head lolled. It surprised her she had the fortitude to right herself.

"They can't cut the program." His voice revealed the fierce protectiveness she'd come to count on, one that sometimes got in the way, truth be told.

"Closed session meeting last night sealed it." Her coffee burned its way down her esophagus.

"Is that even legal? To schedule a school board meeting on the night of your spring concert?"

"I don't know if it was legal, but I'm convinced it was intentional. The whole community showed up at the concert, not the meeting. It's a wonder they had a quorum for the vote. Not that a small thing like regulations could stop a wrecking ball like Evelyn Schindler."

"That's one—"

"Watch your language."

"—*driven* school board president."

Lucy's sigh started at her toes and worked its way upward. "A skilled manipulator of thought."

"Or lack of it."

She almost smiled at his assessment. Would have, if it hadn't been such a tragedy. "When I think of what they were plotting while my students were singing their hearts out..."

"I wasn't the only grown man shedding—what do you call it?—*tender* tears. Your students' music moved us. In a good way. Your best concert ever, LucyMyLight."

Strains of the concert's high points replayed in her bulging brain, soothing and aching at the same time. "My *last* ever, Charlie. My last ever."

The day the music died.

2

"Want to go for a walk?" Charlie sounded like a new dad trying to figure out how to make his toddler stop crying.

"No."

"Retail therapy?"

Lucy considered aiming her mouthful of coffee just slightly in his direction. She swallowed. "Retail therapy? Where did you even learn that term?"

"*Live... with Kelly and Michael.*"

"You watch too much daytime TV."

"Nothing else to do. I only watch while I'm loading the dishwasher."

In a second-fragment, she flew through an image of his typical day since taking early retirement from the paper mill a year ago. *Retiring.* The word sat like lethargic rocks in her stomach. Some cultural *advances* should never have become a staple of the American dream, in her opinion. Like red donkeys versus blue elephants—or was it the other way around?—she and Charlie might forever disagree on that point. He'd been on a countdown toward retirement since his first day at the mill thirty-five years ago. She'd resisted all discussion of stepping away from the passion that secondarily happened to provide her a paycheck—steering young people toward an appreciation for music. *I love you, but worms are not a passion, Charlie.*

He disappeared from the kitchen for a minute and returned with a bottle of ibuprofen. Like announcing a cure for cancer, he plopped the bottle on the table in front of her. "You'd probably appreciate a couple of these, huh?"

He's a good man. He's such a good man. I should be grateful to have a husband who doesn't try to stuff chocolate in my mouth as a cure-all for this.

"I know," Charlie said, opening the bottle, dumping two green capsules into his palm, and extending them toward her. "What you really need is chocolate."

Boom.

"We'll go out to Bernie's for broasted chicken, and then we'll split the chocolate volcano for dessert. With vanilla bean ice cream. *And* whipped cream."

He looked so hopeful this would be the peace offering she'd embrace. Fix-It Man strikes again. Rescues brokenhearted damsel and gets *his* favorite meal in one fell swoop.

"LucyMyLight?"

She pushed out her diaphragm as if preparing for a high note. *Hold. Hold. Exhale.* "I think I'll go take a shower. If you want to bring home chicken from Bernie's, that's fine. I don't feel like being out in public right now, though. Okay?"

"Rumors are flying, huh?"

She hadn't thought of that. But yes, they probably were. The budget conscious would cheer the school board's decision. All the smart people would be in an uproar. Did she just think that completely judgmental thought? Yes, she did. Today wasn't the day to work on a better attitude. Today was the day to spend an inordinate amount of time in the shower, the music in the bathroom cranked full blast, like the water, and later succumb to a chocolate coma. She'd beat back chocolate guilt with a fire poker if necessary. Two. One in each hand.

"Help has arrived," Charlie announced, rustling thin plastic and thunking around the kitchen.

Lucy sat on the front edge of the couch, planted her palms on her thighs, and tried to stand. Her second attempt succeeded. *Halfhearted effort accomplishes nothing.* She had a poster in the music room that confirmed it.

Charlie looked up from where he'd laid out their supper on the granite island. "Oh no."

"What?"

"It's that bad, huh? I haven't seen that sad-looking sweatshirt since the stretch when Sam wasn't sleeping through the night."

"You mean the eighties?"

"Stubborn son of yours."

"Ours." She paused. "Mostly yours." She picked at the loose, crispy skin on a chicken thigh. "Did you bring—?"

"Yes." Charlie smiled.

"Mashed potatoes?"

Crestfallen. There was a word for the expression his face made. "No. I thought...under the circumstances...you'd want fries with that."

With silent apologies to all the hardworking people whose job it is to ask, "Want fries with that?" Lucy let her mind drift to the suffocating smell of overused cooking oil embedded in the fabric of her color-defying sweatshirt. Not that a burger joint would let her wear her own clothes to a new job. She'd be assigned a uniform. Something in a shade not even close to complimenting her skin tones.

"Lucy? Are fries going to be okay? I really don't want to go back." He sliced along the chicken's sternum and pulled off a hunk of white meat.

She leaned her elbows on the table. "You're too good to me."

"Well, yes. That's a given." He winked.

"That you would even consider going back for mashed potatoes..." She thought the hot shower had pelted all the tears out of her. But no. They had friends.

"Hey, hey, hey, Luce. I'm not that wonderful." Charlie moved to stand behind her chair and started a boxing manager's version of shoulder massage. The pressure he applied showed it was to relieve his own tension, not hers. Bless him.

She sniffed back tears and patted his hand on her shoulder. "I'll be okay. Just not right now."

"You'll find another job." His imitation of Perky Life Coach fell flat, like two-week-old amorphous road kill.

"It wasn't a job. It was my life's work."

The RIF letter had probably eaten through her purse like acid by now. She loved that purse. Chicken, she did not love. Not tonight. She'd lost her taste for chocolate, too. That couldn't last long without dire consequences.

She'd been riffed. The term stabbed like a cross between *fired* and *assassinated*.

RIF. It once meant *Reading is fundamental*. An excellent thought. *Reduction in force* turned the acronym sour, biting. RIF. Could she live the rest of her life avoiding those three letters?

The name *Charlie* used two of them. Not that he could help it.

The worst Monday in the history of Mondays. Perhaps in the history of days.

She'd slogged through the weekend, dodging despair, awkward questions, and piles of bitterness like tiptoeing through a heavily used dog park. Avoiding church seemed counterproductive, though tempting. But arriving late and leaving early kept it from turning into a sympathy-fest, a party she couldn't handle yet. Too soon.

Then dawned the inevitable Monday. Lucy didn't have to wonder who among her teacher friends had heard the news and who hadn't. Their eyebrows told the story. Neutral eyebrows? Hadn't yet heard. Pinched together with a slight head tilt—even without a word spoken—signaled a wave of sympathy that became an undercurrent riptide by midday. Survivor guilt kept some from talking about it. They'd received contracts, not RIF letters. Others expressed their sympathy in rib-dislocating hugs. Like Charlie's.

If it hadn't been frowned upon by the administration, she would have stayed in her room to eat lunch. Lucy considered breaking the rule this one time. What was the worst they could do? Fire her? It was the closest she'd come to laughing in days.

But she'd vowed to behave herself to the end. Two more weeks of classes. No vindictive actions. No tantrum-like rebellion. No cement in the toilets or graffiti on the walls. No letters to the op-ed page of the local newspaper. No anti-school-board picketing or bucking-the-establishment T-shirts. No coasting. No phoning it in or phoning in sick.

"Résumé polished and ready to send out?" Ania Brooks slid onto the one blank square foot of Lucy's desk.

"At my age? Your chances of finding another teaching position are a lot better than mine. Age does make a difference." Lucy snagged a piece of music too near the clear zone. "And right now, applying to substitute teach feels like the difference between running a karaoke machine and composing a symphony." The words felt coarse in her mouth. Some of her favorite teacher friends subbed. Bright, skilled educators willing to rewrite their schedules when needed. What was wrong with her?

Ania flipped her thick, loose black braid over her shoulder. "Don't be so sure. About my *flood* of opportunities." The younger woman picked at loose threads of her fashionably tattered jeans.

Wait. Tattered jeans? Not exactly school policy for staff. Lucy retrieved her insulated musical score lunch bag from the bottom drawer of her desk and pointed with it toward the spot where Ania's bare knee showed through. "Dress code no longer applies to you?" She feigned the voice Principal Rust might use.

"*Contract* no longer applies to me." Ania pulled an apple and a bag of microwave popcorn from the deep pockets of her hand knit cotton sweater. Coral. Somehow the faded blue scarf looped around her neck and the fused glass pendant strung on what looked like an athletic shoelace screamed "art teacher" without Ania having to wear a nametag.

"We're still under contract for two more weeks."

"A technicality, my rules-bound friend. Merely a technicality."

The hitch in Ania's voice belied her mask of courage. She couldn't be as cavalier on the inside as she appeared on the outside. Not yet thirty, she was sure to find another position, though. Maybe in a larger public school without the budget woes of Willowcrest.

A school without budget woes. And other fairy tales.

The two recently riffed walked the hallway toward the teachers' lounge as they had many times. Never this speechlessly.

Was it imagination, or did the level of chaos in the halls decrescendo as they passed? Like freeway drivers reducing speed for the quarter mile before and quarter mile after the state patrol car parked in the median, the students quieted a few decibels then resumed their normal ear-splitting volume.

"Hey, Mrs. Tuttle. Sorry to hear about—"

A rib-jab cut the condolences short. "You tosser!" the jabbing student said.

Ania's look revealed a need for translation.

"A meme of British expressions is circulating on social media again. The students think we don't know."

Ania smiled. "Some of us don't. What's a 'tosser'?"

Lucy lowered her voice. "Idiot."

"Are you going to call the kid on it?"

"And blow our cover? Don't worry. I gave him my fierce look."

Ania's laughter helped cut through three days' worth of tension. "On you, Lucy, *fierce* looks an awful lot like 'Oh, you sweet child.'"

"I squinted." Even Lucy knew that was a lame defense. She *squinted*? No wonder the school board had no trouble making her a target.

She stopped herself. Paranoia wouldn't help. They'd dumped both the arts and music *programs*. Nothing personal, they'd said. Uh huh.

Two steps into the teacher's lounge and Lucy knew the better choice would have been to sneak her egg salad and grapes in her room. The lounge erupted with anger over the school board's decision. Lucy wasn't ready for anger yet. She held tight to despair.

Her students watched her more intently than on an ordinary day, far more intently than they would under normal circumstances this close to the end of the school year.

She'd taught them to watch her hands and facial expressions as she directed, to listen for the breaths she took that reminded them when to breathe. She'd taught them to express the emotion of the song, not how they felt about the person with whom they shared a music folder. Under her guidance, they'd learned to focus on what the music asked of them, often a response contrary to their young nature, their personality, their mood, or how much sugar they'd ingested at lunch.

Now they watched, too, for her reactions to a crashing tympani blow to her life's plans.

Lucy thought she had an easy out for the day. A way to survive without having to engage her overworked brain. Each class could review the video of last week's concert. Tradition. The students expected it. She, on the other hand, didn't anticipate the fortitude it would take to sit through that many replays of her final moments in concert.

The nuances she witnessed on the video hit like memories of a too-recently deceased loved one. Would she have felt the same if her choirs and band hadn't performed as well as they did, if the audio didn't resonate now as near-perfection with enough sniffles and coughs and squeaks for her to know it was real?

The parallels to life made her jaw hurt.

Everything seemed too tender to touch—a deep, aching life bruise. Memories of the concert high points, career high points, the tears when especially sensitive young people caught a whiff of gossip about the music and art programs, the condolences of other teachers, the pile of unfinished projects on her desk...

The music-themed gifts from students—mugs ad infinitum, pens, coasters, pins, note cards. Handwritten thank-you notes tacked to the corkboard, representing a hundred more in file folders.

She mattered here. The music mattered. It changed people. Including her.

If her heart kept beating until the end of the school day, she'd start packing her personal belongings. Taking them home a few at a time would be less painful, wouldn't it? You can't rip a bandage off a wound that's still bleeding.

"Are you still here?" Ania's voice carried too well in the acoustics of the empty, high-ceilinged music room.

Lucy nestled another resource book—*Music and the Young Mind*—into the nearly full cardboard box on the seat of her office chair. She reached for another from the top shelf. "Just a few more minutes."

Ania's clogs clunked across the tiled floor toward the narrow office. "There's no such thing as overtime in this business. Or so I've heard," she said, her sarcasm more pronounced than ever.

"Never was in it for the overtime."

Ania's floor gazing told Lucy the young woman had more responses than anger in her.

Another book landed in the box. "Did your afternoon go okay?"

"Did yours?"

"No." Lucy flipped open the front cover of the book in her hand. "Ah. Thought so."

"What?"

"I thought I'd purchased this one. But I've been here so long, I had to check to make sure it wasn't school property."

Ania planted her hands on her hips. "After what they did to us, you're worried about accidentally absconding with one of their books?"

"I want to do this right. Have to do this right."

"Faith getting in the way of reason again?" Ania flicked the edge of the verse-of-the-day flip calendar on Lucy's desk.

"They're not mutually exclusive, my friend." Lucy deposited another couple of books in the box, enough to reach both the box's and her back's limit. She'd been looking forward to her summer pace of exercise. For once, September wouldn't change her daily schedule. Some people in her position would be grateful. She couldn't imagine mustering *grateful* yet. Ever.

As expected, Ania let the faith conversation drop. She sighed with her eyebrows, shoulders, and lungs, then turned and called over her shoulder, "See you tomorrow. If I decide to show up."

Lucy gripped the back of her office chair. Why hadn't she thought of it before? She could wheel the box-laden chair out to her car. She could move a lot more in one trip that way.

Except for the doorways she had to maneuver, the idea turned out to be genius. She loaded the rear seat of the car and wheeled the chair back into the school, down the hall, around the corner, and into the music room.

It wasn't empty.

"Evelyn?"

"Lucy. We...we thought you were gone for the day."

"The light's on in my office." Maybe it was time to consider her doctor's suggestion about blood pressure medicine.

Evelyn Schindler glanced at the two men with her, as if they could offer a response on her behalf. "This is a team from our contractor's office," she said. "Just here to take some measurements. We'll"—she turned then to face Lucy—"try not to get in your way." The woman's smile hadn't had much practice in her seven decades of living. It was so rusty now, it almost creaked.

"Measurements for what?" Lucy would ask forgiveness later for noting that Evelyn's posture shift made her look like a sandhill crane prepared for liftoff.

Evelyn turned toward one of the men who extended his palm as if to say, "This is all yours, lady."

"We hope to have...the room converted into two regular classrooms before school starts in the fall. Getting estimates now." Matter of fact. Matter of farce.

And that's when the rickety bridge Lucy's emotions had been teetering on splintered. Lucy spiraled—free-falling without a chute—into the blind, bottomless abyss. The school board decision hadn't merely eliminated her job. It had obliterated music. Remodeling shouldn't have come as a surprise. Contractors were inevitable, contractors hired to erase the evidence that music once lived in this space.

As she plunged deeper into the cold darkness, she heard her insides crying over the loss, crying for the children.

3

Maybe it hadn't been such a good idea to strip her office and the music room of her personal belongings before the end of the school year. Her administrator thought she'd find it comforting that school-owned equipment, instruments, music stands, and textbooks would be sold or given to another school. Comforting? Like iron spikes pounded into her vocal cords.

It took more willpower than she'd ever used against a sleeve of Oreos not to respond, "You mean, sold to a school that cares about the arts?"

Even the inattentive students noticed how bare the room seemed with the life stripped out of it. But their enthusiasm for the upcoming summer break overrode somber moments. Lucy had spent the last few days of the school year—of her career at the school—encouraging students to consider other ways to keep the music going. Private lessons. Small ensembles. Singing or playing for church activities. Community youth music programs, which an especially sharp student was quick to point out didn't exist in Willowcrest.

"You, Mrs. Tuttle. You could make a program."

In the community that allowed its school board to evict music? Not an environment likely to get behind a project like that.

"I don't know what my plans are right now," she'd answered. That was no fabrication.

For the first time in twenty years, she had no plan. Even in the years before she'd applied for this position, she'd had a plan. She'd get Sam and Olivia through school, taking every recertification necessary and every continuing education class she could fit in and afford, serve for every music boosters event, accompany the choirs, sing for every wedding, play for every funeral, stay in the forefront of hiring minds, so when her dad was ready to retire, she could be considered a logical choice to take over the program he'd crafted, despite her years off to raise her family. Her heart soared in each of those endeavors. They brought significance and meaning while she waited for her life's goal.

Her father's music program left a true legacy in the community. He died from a brain aneurysm four years before he was slated to retire. Lucy's plan got bumped up. But the decision-makers agreed they didn't need to look farther than Cottonwood Street for a replacement for Lucy's beloved father.

Her final note in the final measures of her role as music teacher for K-8 at Willowcrest School—a bitingly cold thought on an otherwise cloudless day in early June—was that she'd failed her father.

She'd grown hoarse with the effort, but in the end had been ineffective in convincing the new decision makers that art and music weren't just *as* important as science, math, English. They helped students understand and perform better in academics. Stay in school. Develop a success mind-set. Become more well-rounded young citizens and learn to work in community. Music gave students interests and options.

She couldn't stop the bulldozer of budget cuts that demolished her father's programs. It had run over her too.

How much therapy would it take to recover from that?

It couldn't be the flu. The nausea that hit as she pulled into the driveway at home came on too quickly for it to be the work of a virus. It had all the earmarks of melodramatic disappointment.

For nineteen summers she'd skipped across the threshold, relatively content the school year ended, tamping down her exuberance for the following September with the natural exhaustion from nine months of intense labor and the late-in-the-school-year squirreliness of students more than ready for a break.

She steadied herself against the car door and waited for the wave of nausea to pass. It didn't. She'd have to cross the threshold despite it.

Charlie had the door open before she got there. Again. "Hey, LucyMyLight. Welcome home." He grabbed one of the boxes she carried and stepped aside to let her in.

Nausea *and* headache. The house bulged with people.

Congratulations! Way to go! You should be proud of what you've accomplished! And the stinger—*Happy Retirement!*—twirled in a cacophony of well-intentioned noise from friends and family. Family. SamWise and Olivia?

"Hi, Mom." Her daughter leaned close to Lucy's ear. "We tried to talk Dad out of it, but he insisted you needed a party."

"I so don't," Lucy whispered back without moving her faux-smiling lips.

"I know. I'll help you get through it."

"Sam, honey. How did you get off work on a weekday?" Lucy hugged her taller-and-thinner-than-Charlie son. Leaning into his sturdiness steadied her.

Charlie joined the hug fest. "We pulled it off, didn't we, kids? Okay, let's get this party started. Happy Retirement Day, Lucy." He cupped her face in his hands. In a volume only she could hear, he said, "Making the best of it, love." His kiss on her cheek came first. Then the tenderest of smiles.

If it hadn't been for that brief hint of understanding, that grace note, she might have ground her teeth to nubs. Locked herself in the bathroom. Called 9-1-1 to have them break up the hilarity on Cottonwood Street.

Instead, she braced herself for the well-wishes and the so-sorry-to-hears and the you'll-find-something-else-to-dos. The buffet Charlie organized consisted of all desserts. Smart man.

No, he didn't! Yes, he did. Presents. He'd let people bring presents.

Lucy caught Olivia by the arm. "I'm not a drinker, right?"

Olivia smiled, every penny of her multi-thousand-dollar orthodontics worth it. "No. You're not."

"Just checking."

"Hang in there, Mom. I'll help you write thank-you notes."

"Giving birth to you was the best idea I ever had."

Olivia's laughter lifted some of the awkwardness of opening gifts that felt like funeral party favors. More music mugs. Miniature golf clubs in case she found, as the card said, "a little time for golf." A framed photo of Lucy, arms raised, directing her Christmas choir. "See there?" the giver noted. "It's a clock face. Get it? Time on her *hands*?"

After two hours of that, SamWise wrapped his arm around his dad's shoulder and told the crowd, "Hey, gang. Hate to break this up, but the family has dinner reservations tonight. So...thanks for coming. We know Mom has appreciated your support over the years, and—"

Olivia stepped in. "—and so do we. Don't worry about helping with clean up. We'll take care of that. Mom has longed for the day when her kids offered to clean up without being asked." A smattering of guffaws served as a benediction. "Thanks so much, folks."

Charlie helped steer well-wishers toward the door. When the last of them had exited, to waiting cars parked who knew how far down the block in order to sufficiently surprise Lucy when she'd walked in, Charlie closed the door, leaned his back against it, and said, "There. That helped, didn't it? Nice party to kick off your newfound freedom?"

Freedom? Of all the words in all the world, the only one that came to her mind was a social media British word she couldn't voice.

Dinner reservations were at their own kitchen table. Delivery pizza. Sam's idea. *Good son. Good son.*

Charlie picked the last bit of Italian sausage from one of two pizza boxes while Lucy reached for the final mushroom. "So," he said, "are you ready for..."

Don't say it. Don't say it, please. No, I'm not ready for the first day of the rest of my life!

"...the *rest* day of the *first* of your life?"

"What?" Her voice was harsher than she meant it to be. Had he twisted the phrase on purpose?

"More root beer, anyone?"

A Tuttle tradition. Pizza and root beer. Now that she thought about it, it sounded more disgusting than quirky.

"I'm switching to water," Lucy said.

Charlie whisked Sam away to show him the website on which he'd found the best prices for worm farm supplies. Sam's face looked like tolerance with a thin mask of interest.

Olivia crunched the empty pizza box as small as she could and stuffed it into the recycling bin. "Mom?"

"Hmm?" Lucy gathered paper plates and followed on Olivia's heels.

"You hadn't prewarned Dad that a retirement party was the last thing you'd want?"

Lucy opened the door of the dishwasher and loaded silverware. "It didn't cross my mind." She paused to rinse the pizza cutter. "Sometimes husbands think they're helping when they're not." She eyed her should-have-gotten-her-PhD-in-psychology daughter. "In his mind, he was doing something thoughtful." *Thoughtful.*

She dropped another fork into the dishwasher basket. "I love him. Sometimes that has to be enough."

Olivia searched for storage containers for the leftover desserts. "Do you have a lid for this one?" She held a rectangular container aloft.

"Bottom drawer near the sink. Toward the back."

"Found it."

Lucy closed the dishwasher and leaned against it. "Olivia, why didn't *you* succeed in stopping your father?"

Her daughter turned, her eyes wide.

"Stop me from what?" Charlie asked, poking his head into the kitchen.

"Girl talk," Lucy answered, drawing a smile from where it had been hiding. "What did you need?"

"Internet's out."

"Again? I'll be so glad when they finish laying the underground fiber optics."

"Where do we keep those big legal pads of paper? I need to sketch out my plans."

Don't ask me. I'm the planless one. "The closet in the office. Second shelf from the top. Right-hand side. Behind the—"

"I'll find them. Thanks." His head disappeared, leaving the women alone again.

Olivia's eye roll joined her pert smile, the equivalent of a silent "Sure, he will." Lucy counted down. *Ten, nine, eight, seven, six, five...*

Charlie reappeared. "Where?"

Lucy started the dishwasher with a practiced sigh. "I'll show you."

Olivia snapped the lid on another storage container of leftover desserts. "That's okay, Mom. I'll show him." Then mouthed for Lucy's eyes only, "Because I love him."

The room emptied. Life had emptied. Now what? *Tone down the drama, Lucy. Life isn't over. It just feels like it.*

She was too young to consider this stage retirement, too old to consider it a serendipitous opportunity to retool. But...

From this angle, one thing became clear. Life had just handed her time to repaint the kitchen. And it needed it. What else had she neglected?

"LucyMyLight?"

The man who loved her enough to throw her a retirement party when she'd been riffed. The man who committed his forevers to her. The man who probably had no idea how his surprise had drained her. Or that he had pizza sauce on his chin. "Yes?" She swiped at a water spot on the faucet.

"Do you know what I think we should do?"

She waited. Buy an RV and travel all fifty states—well, the forty-nine accessible by RV, work at odd jobs along the way to pay for gas and teach music lessons from aluminum lawn chairs? Let the worms have the house and they'd move into the garage? Move to Nashville or New York where people care about music?

"I think we should invest in a better computer. I'll be taking orders internationally, I imagine."

"For worms." So that's what deadpan sounded like. She couldn't retract the echo.

"Are you okay?" Charlie looked offended. *Charlie* looked offended.

She skirted around him and headed down the hall to the bedroom. "Okay? Not. Even. Close."

4

How long had it been since she screamed into a pillow? Decades. Probably when SamWise burst from her loins pre-equipped with colic. Today it only took one muffled scream before she stopped herself. Bad for the vocal cords.

As if it mattered anymore.

The bedroom door clicked open. Lucy sat on the edge of the bed, the pillow in her lap rather than over her face.

"Hey." Charlie lowered himself to sit beside her. She felt her body tilting toward him and grabbed a hank of the comforter to keep herself upright.

"Hey," he said again, reaching for her knee. The pillow stopped him. She moved it to the head of the bed, where it belonged.

Don't make me talk, Charlie. Please don't make me try to explain what I'm feeling right now.

"Do you want to talk about it? You can't fix what you don't acknowledge." He rubbed her knee as if it were the seat of her emotions. "See?" he said. "I'm a better man for watching Dr. Phil."

What did it mean when a line she would have found funny any other day smelled as bad as paper mill emissions? She had no sanity comforter to grab to keep her from sliding farther.

Wasn't that supposed to be one of the perks of faith? Not falling apart when your world does? Blindly assuming it will all turn out fine one of these days? Sanity glue?

"'I will not say, "do not weep," for not all tears are an evil.'"

What language was he talking now?

Charlie lifted his hands, palms up, as if she should have instantly understood him. "*Lord of the Rings?*"

"I thought you slept through that."

"Not all of it." Those gray-green eyes she'd fallen in love with her freshman year at LaCrosse, almost teal today because of the shirt he wore—her favorite, she noted—brushed over her with a penetrating look that said, "I don't understand any of this, but I want to."

She rested her head on his shoulder. It's then she noticed. *The bedroom needs paint, too.* Lucy didn't move but scrunched her face against the sting of her own indictment. Random thoughts dribbled over into the spot reserved for rests, like the work of an undisciplined musician. She cleaned up the dribbles and leaned into the act of resting her head on his shoulder. Breathing. Drawing strength from his. Filling the rest-space with what it required—silence.

Silence that didn't stop counting beats.

She'd made it part of every class's curriculum—a discussion about rests. Their value. Their weight. Their purpose. "They're not a vacuum of nothingness," she'd said. "They have meaning. The music isn't the same without them. Rests deserve as much attention as the notes you sing or play."

Without breaks in the sound, she told them, music has no pattern or shape.

"Don't lose your intensity. Don't lose your focus. You can't afford to ignore rests. Honor each rest as a precise note of soundlessness. Active silence."

She heard her mind rehearsing the instruction she'd given young musicians for almost twenty years. Some listened better than others.

As she *rested* her head on Charlie's shoulder, a knot in her neck untangled itself. Slowly. Like air leaking from a balloon. She felt the knot loosen its grip on her bit by bit. Lucy remained as still as she could, fearful of disturbing the fragile peace.

How many minutes, how many heartbeats, how many rests between beats did they remain that way? How many more, if Olivia hadn't tapped at the door. "Mom? Dad? Sam and I are going to a movie. Want to come?"

Without even asking what was playing, without knowing if it was a movie with any worth at all, Charlie answered for them. "Yes! Great idea. Come on, Lucy. That's just what you need."

The knot in her neck hadn't disappeared. It relocated. To her throat. She couldn't swallow her own saliva.

Charlie stood now, tugging her to her leaden feet. The door behind him opened. Olivia peeked around her father.

The war raged as real as the *Lord of the Rings* final battle. Acquiesce? Or say what she really thought? Go along, against her inclination? Or stay, against theirs? "I'm staying home tonight. You go on ahead. Have fun."

Charlie's dumbfounded expression was getting a workout. "What? No. Come with us, LucyMyLight."

"Staying home." Lucy shook out of his tugging grip and ducked around the end of the bed toward the en suite bathroom. "I think I'll take a shower, crawl into my jammies, and read a book." *Or something.* "But don't let me stop you three. Enjoy the movie."

Olivia's eyes asked if she meant it. Lucy nodded. "Go." She shut the door to the bathroom before anyone could raise an objection. One of them must have heard when she depressed the lock button, leaned against the door, and slid to the cold tile floor. Probably Olivia. Lucy heard her magnificent daughter steer the boys into the hallway as if it were no big deal that her mother chose to stay home from a family outing, no big deal that she wasn't handling the simplest things well, no big deal that the woman who normally went along with the crowd without voicing an opinion had turned vocal. A little bit vocal.

She stayed in the shower until the hot water heater protested. Towel snugged around her, she fingered volumizing mousse through her hair. Her natural curls behaved themselves well with the new—expensive—mousse Olivia had recommended. "How much longer will you be mine?" she asked the can. "Not sure you'll work into the budget anymore."

Lucy wouldn't start that list tonight. All the things they wouldn't be able to afford without her full salary. Private, nontraditional school. Nontraditional pension. Not the night to think about that.

In her pajamas, she padded through the empty house to the kitchen, noting along the way that the house breathed again, and it wasn't just the air conditioner. The closed-in feeling it had sported for days was gone. Doorways resumed their normal width. Hallway walls no longer pulsed closer, closer. Airy, light, a home simple enough to keep the taxes lower than the fancier houses on the block but large enough to meet their needs and then some. With the mortgage paid off—thanks to the nineteen better years with full-time incomes for both of them—they might have to cut some other corners but should be able to keep the house. Charlie couldn't seriously think they'd need to get rid of her car, though, could he?

We'll go places together, Lucy. Why would we need two cars?

We can't be together all the time, Charlie. We can't.

No. Absolutely can't.

What time was it? She glanced at the clock on the microwave. Time enough to sit on the deck for a while before the movie crowd dispersed. If the bugs weren't too bad.

They were. Lucy took her book—a best seller she'd never heard of but Ania recommended—and retired to the kitchen nook so she could still have a view of the backyard, the paths lit by solar-powered patio lights.

The nook felt like a glorified window seat, the wide curve allowing an upholstered bench along the window wall with a round antique table and two chairs opposite the window wall. Charlie hadn't griped about the feminine décor. His favorite spot was the breakfast bar. Hers was nestled into the blue and white toile with the mix of yellow patterned pillows.

She propped herself with her back against one of the side shelving units—all white, like the cupboards—feet crossed at the ankles on the toile. She reached behind the pillows to retrieve a lightweight quilt. The air-conditioning felt good on her face but cool on her arms and feet. Wrapped and settled, she opened the book and waited for the words to make sense, for her world to make sense.

The words swam on a sea of "How can this be happening?" After several failed attempts at page one, she closed the book and stared into the dark.

Seasonal Affective Disorder. That's what she had. An inability to function well in a season when thick clouds formed by budget-nervous school boards block out the sun and make life-altering decisions. The medical community could offer no lamp for that kind of S.A.D., no uniquely designed light-emitting apparatus that would make a difference.

Light-emitting apparatus. That's what she needed.

"Don't mock me," she said to the barely visible stepping stone illuminated at the head of the garden path. She couldn't read the inscription from this distance, but she knew it by heart:

> *Your word is a lamp before my feet*
> *and a light for my journey.*
> —Psalm 119:105

"Is it wrong to allow myself to be miserable for a while?"

The words scared her. Two months ago she would never have imagined making a statement like that. The RIF decision had changed her into someone even she wouldn't want to be around.

5

"Good. You're home."

Ania sounded relieved. Lucy stared at her cell phone. Where else would she be on a jobless weekday morning in the summer?

Summer. Lucy used to call it summer *break*. Not a season, but a half-rest between school years. "Where are you calling from?"

"The library."

"That would explain the clandestine whisper voice." Lucy mimicked her friend's tone.

"I think I may have discovered a way we can sue the school board."

"Ania, we're not suing the school board." Lucy fingered the tiny copper hoop earrings she wore—another gift from Ania. She glanced out the laundry room window at the thermometer near a hanging basket of flowers. That warm already?

"We wouldn't get very far suing for what they did. These RIFs are happening all over, sad to say. But we might be able to get them for how they went about it."

Lucy closed the lid of the washing machine and punched the code that started it filling with water. Nothing happened. She bumped a spot to the right of the control panel with her fist. Success. "'Get them?' I don't have any desire to exact revenge." Not that she hadn't prayed God would.

"Sure you do. Don't let your sadness override your anger, Luce. We need to take action. Can you come over to the library while I'm still here?"

It took all of three seconds to rehearse her plans for the day. But poring over legalese wouldn't improve on that schedule. "I'm stuck here until after eleven at the earliest. Charlie has my car. His is in the muffler shop." Yes, it was only a mile or so on foot—each way—but...

"This is just preliminary anyway. I'll e-mail you links to the information I found."

"I don't want to even think about taking legal action, Ania. Really."

"You'll change your mind when reality kicks in and you have to start waiting tables at Bernie's."

The idea lodged in her throat like a crosswise potato chip—not likely to kill her, but highly unpleasant, considering. She'd worked at Bernie's during summer breaks from college. Had her career path come full circle?

Lucy returned to the task of taping around the windows, doors, and cupboards in her kitchen. She'd start painting when Charlie returned from his trip to the home improvement store. Picking out a color online changed the whole "send husband to the store for paint" challenge.

She hoped.

Ania had asked one more question before hanging up. Had Lucy picked her poison yet? Ania's poison of choice? Cheetos. Crunched one at a time, in rabbit bites. A tension-easer, she claimed.

Lucy's choice? She stopped pouring coffee into the white porcelain mug with an inset for a stack of Oreos. "Developing a new food addiction is going to help how?" she'd asked Ania, pulling a different mug from the shelf that held her collection. Something plain, with no handy compartments for Oreos, caramels, or Cheetos. Lucy dumped coffee from the first mug into the second.

She would have been proud of herself, but she couldn't get past mourning the missing cookies.

The bright blue painter's tape clashed with the faded, nondescript blue on the kitchen walls. From a distance, it looked as if she'd trimmed her kitchen in Ugly. This stage—the ugly stage, the *in-between*—always lasted longer than she hoped. Befores and afters hold merit. A "was" full of memories and a "future" full of awe. But this middle? When nothing looked or functioned as it should? Lucy fought agitation with every strip of blue tape she positioned.

She could strip off the white crown molding at the ceiling and risk splintering one of the pieces, as Charlie had the last time they'd painted. Or she could climb the ladder and tape the entire perimeter of the room. She opted for the latter—and the ladder.

Fifteen minutes into the process, her neck and shoulders threatened mutiny. She allowed them a short break, during which she refilled her disturbingly plain coffee mug and clicked the remote to start her playlist of music.

Bad idea. Bad, bad idea.

Lucy didn't hear professional musicians, auto-tuned and digitally mixed. Her mind heard children's voices, children's talent on the instruments. Children discovering the power of music to change things. Children finding expression, coming out of themselves to care about the community of sound, pushing themselves past what they thought unconquerable limits, discovering not only a safe but soul-enriching outlet for their emotions.

She abandoned the ladder and grabbed a stack of Oreos.

"Just a few," she told the room trimmed in Ugly. She resealed the cookie bag and tucked it deeper into the pantry. As if that would prevent a return trip.

The hated stage—between before and after. Music swelled around her, a soundtrack for her mood. Ladder height made her dizzy. It had nothing to do with the sugar rush. She switched from coffee to iced tea and retreated to the deck to wait for Charlie to return with paint.

The sun hadn't yet climbed over the roof and the maple—a formidable team that eliminated Lucy's need to open the deck table umbrella. She'd left the music on in the house. Filtered through walls and windows, it seemed less threatening to her sanity.

The swivel-rocker patio chair welcomed her presence as if it had been waiting for someone to recognize what a beautiful day it was. Above the decibel level of the filtered music, she heard the thrum of hummingbird wings. The bird darted in and out of the fuchsia Million Bells' tiny, petunia-like blossoms.

The hanging basket offered a bright spot in a still-getting-a-foothold garden. Winter stayed too long. Didn't it always? The herbs and smattering of vegetables she'd managed to plant between end-of-the-school-year activities survived under a layer of old sheets on frost warning nights. The hanging plants filled her kitchen counters—a makeshift greenhouse that necessitated Chinese take-out for dinner.

Charlie didn't complain. Except about the lady bugs that hitchhiked on the underside of leaves and stayed indoors after the plants were returned to their spots on and around the deck. With Charlie, even complaining took on an air of amusement.

Lucy's blessed time alone hadn't exhausted itself before Charlie joined her on the deck.

"Isn't it great to have freedom to enjoy a morning like this?" He slapped a gallon of paint onto the patio table.

"On days when I didn't schedule summer music lessons, I could have mornings like this every year."

He removed his ball cap, swiped at his forehead with the back of his hand, and resettled the cap. "But it's different this year. *Endless* summer, LucyMyLight. No need to spend June, July, and August getting ready for September. No school-related summer lessons. No interrupted plans so you can drive to the school for...whatever."

She should have found some sort of comfort in his enthusiasm or in the truth of what he said. Instead, she focused on his use of the word *interrupted*. Whose plans? She hadn't considered it an interruption. Charlie's work at the paper mill kept him busier than ever during the summer, until he retired. Why wasn't that considered an interruption?

"Trying to look on the bright side," he said, picking up the paint can and heading for the kitchen door.

Your bright side is blinding me, Charlie. "Thanks for getting the paint."

"If you ask me, it looks like what you already have on the walls."

"It isn't. A shade darker."

Charlie turned to face her. "Darker? We need darker? I thought you liked all the light in there."

"I do. But with the white ceiling, cupboards, and trim, there wasn't enough contrast. You'll see. It'll be stunning. Ania thinks so too."

Gravity pulled Charlie's facial features south. "Ania the Angry Artist?"

"How many Anias do you know? Yes. The Angry Artist."

His mouth twitched. "Do you think it's wise to take advice from her?"

This is how daily conversations would go, living with a man with too much time on his hands and no clue how offensive his words could be? "You're not telling me who I can have as a friend, are you?"

"And sign my own death certificate? No." His chuckle showed obvious comedic intent. Then his facial expression changed. "Just saying that her anger is... toxic. And your emotional immune system is compromised."

"Dr. Phil?"

"Great episode." He stepped into the kitchen, then opened the door again to call to her, "Hey, you mind if I turn off the music? Or turn it down? My earlobes are bleeding."

Eardrums, and no they aren't. She followed him into the kitchen. "Go ahead. I was thinking of heading to the library for a while, now that the car's back."

The music crash-landed. "Okay." He craned his neck, surveying the mess in the kitchen. "Will you be back in time for lunch?"

His question could have meant so many things. *How long do you plan to leave it like this in here? How long will you be gone from me? What are you going to do, and can I come, too? Will you be back in time to make my lunch?* Or, simply, *when will you be back?*

"I have some errands to run. Don't know how long it will take me. Can you make yourself a grilled cheese sandwich for lunch?"

"Sure," he said, his face less than certain. "Don't you want to . . . ?" He let the thought dangle and drew concentric circles in the air while pointing at her chest.

No! I do not want to—

". . . change first?"

She looked at the front of her grungy tee shirt. Paint spatters from a previous project. "I-I planned on it." Lucy eased past him and headed for the bedroom to change into an outfit she could be seen in publicly. She settled on clean jeans, an unspotted tee, and a lightweight jean jacket. Errands. Which of those on the list interested her?

The kitchen could wait. Time was something she now had in abundance. She'd talked about taking the hiking path along the river. Maybe today was the day to cross that item off the list. She chose athletic shoes over sandals for that reason.

"What if the muffler gets done before you come home?" Charlie asked when she kissed him good-bye. "We have to pick up the car together. I'll need a ride to the shop."

Lucy wiggled her cell phone. "You can text me. Or call if it gets close to their closing time. Did you have anything else planned for the day?"

"I thought about seeing if the bluegills were biting. Martin said something about wanting to go fishing."

"Great. Have fun."

"I don't have a car."

"Can Martin pick you up?"

"I suppose so."

"Perfect." Lucy snatched her purse and exited through the front door before she thought too hard about inconveniencing Charlie.

She drove to the library, parked in the lot, but didn't get out of the car. She kept the engine running for the sake of the air conditioning, surprisingly useful on a day when the temps were ideal. The sun. That sun beating down on everything. Baking the car's interior. Lucy directed the top vents to blow directly on her face.

Library patrons strolled in and out of the building. Few people *ran* into or out of a library. Young moms, maybe. With toddlers in tow. In the rain. Libraries are destinations of discovery. A lot like musical pieces, Lucy thought. Those who rush to, through, or out of it miss the whole point.

She shook herself out of philosophy mode and made a decision. A bold decision. Discovering how to sue her former employer sounded even less appealing than it had when Ania told her about it over the phone. How would starting a war benefit her students in any way? Lucy longed for reason to prevail. Running into Ania wouldn't help anything. What was she even doing downtown? Lucy put the car into reverse and backed out of her parking spot.

Into a tan SUV with the same idea.

The jolt sent her heart rate into staccato overdrive. She turned off the engine, unbuckled her seatbelt, and jumped out to assess the damage and meet her victim face-to-face.

"Mrs. Tuttle? Hey, I am so sorry. Are you okay?" The shorts-clad teen girl clutched her stomach.

Lucy put a hand on her former student's shoulder. "I'm fine. Are you?"

Kiersten shook her hands at her sides. "Yeah. Fine. It's how I handle stress."

"You're sure you're okay?"

"Fine. Really. I am so sorry."

"What makes you think it was your fault, Kiersten?" *Careful, Lucy. Don't take more than your share of the blame.* She'd heard that from an insurance commercial once.

"Not directly. But—" She turned to indicate the bent woman approaching from the driver's side. Bent with age, not accident damage, it appeared. Kiersten—summer blond wisps of hair stuck to the sides of her face—stepped between the two older women and spoke to Lucy in soft tones. "She insisted on driving us home. Insisted. I guess respecting your elders has its limits. I am so sorry."

When she was in eighth grade, family responsibilities had forced Kiersten to drop band and chorus. It broke Lucy's heart. And not just because they'd needed her on French horn. The accident now seemed a rude intrusion on Lucy's longing to reconnect with the young woman these years later and find out how she was doing.

Kiersten stepped to the side. "Grandma, I'd like to introduce you to Mrs. Tuttle, my former music teacher." The apology on her face couldn't have been more pronounced.

The older woman toddled to the spot where the two bumpers seemed locked in a wild embrace. "Oh, this isn't good."

"It's not so bad," Lucy insisted, mentally calculating the cost of bodywork added to the new exhaust system Charlie's Traverse was getting in a shop down the street. "A little crinkled."

Kiersten's grandmother pinched her eyes shut, then opened them wide. "Still here. It's real, I guess. Well, if I had a license, it'd be gone now."

"Grandma, you don't have a driver's license? You didn't tell me that." Kiersten's face lost a decade in age with that revelation.

The woman touched an undamaged part of the SUV's bumper. "Kind of a moot point now. This is as nasty as a cat in a lace factory." She looked up into Kiersten's face. "Don't tell your father."

"Grandma, I have to tell him."

The woman's shoulders heaved. "Then you'll have to tell him the whole truth—that I overpowered you and took the wheel against your better judgment."

Kiersten and Lucy laughed at the way Kiersten's grandmother flexed her biceps as she spoke.

"I suppose we need to call the police so they can file a report. And exchange insurance information," Lucy said. Her head throbbed. Her first fender bender, ever, and it had to be with one of her former students. Correction. A former student's rambunctious grandmother.

"I already called it in," Kiersten said, indicating her cell phone.

A small, sympathetic crowd gathered. A library assistant. A friend from church. Patrons who got more than they expected—a good book *and* a show. Finally, the law enforcement officer—also a former student but from a decade earlier—and Lucy's insurance agent, whose office was across the street from the library. Handy, in incidents like this.

Statements taken, truth told, bumpers untangled, the women were freed to leave the scene. Plastic bumpers on both vehicles meant they were left with unsightly holes, but nothing dragging behind them. Drivable, but a little broken.

Lucy's sigh expanded to fill the suffocating interior of her Malibu. "Me, too," she said, patting the dashboard. "Drivable, but broken."

Thank goodness for the accident. Thank God for it. She'd managed to divert every conversation that started with "So sorry about your job" to instead focus on the traffic jam in the library parking lot and the two wounded vehicles at the center of attention. That, and the crumpled grandmother who couldn't stop crying. They made an interesting threesome—the ex-schoolteacher who'd gotten additional practice weathering embarrassment, the young woman whose only sin was giving in to her grandmother's request, and the older woman whose keys had already been taken away but now faced the humiliation of *other* people's keys being taken away from her, too.

Lucy hadn't called Charlie. What could he do with his own car out of commission? She drove to the body shop for an outrageous estimate, then opted to drown her estimate sorrows in taco salad at Bernie's. She'd call Charlie after that. Or go home. Or...

"Raspberry lemonade?" the waitress asked.

"Oh, sure." The humor helped somehow. "Let's make some lemonade out of all of this."

"Pardon me?"

"Lemonade. Yes. Thank you." *And some to go.*

Lucy's two forkfuls of taco salad fought each other in her stomach when Evelyn Schindler walked into the restaurant on a path that would take her right past Lucy's booth. Lucy dug into the salad as if looking for buried treasure under the lettuce.

"Lucy!"

Foiled again. "Mrs. Schindler."

Evelyn Schindler sent her lunch companion to a table nearby with instructions to order her a lemonade. What right did *she* have to drink lemonade? Lucy swallowed the hooked barbs of wounded pride.

"I'd hoped to have a chance to talk to you about the budget cuts," the woman said, leaning her Stella McCartney fragrance-of-the-month into Lucy's personal space. "Nothing personal in all that. Purely a financial decision. We had no choice."

Lucy's three-point, fourteen-page rebuttal fell in line like iron shavings to a magnet. How could ripping music and art from the lives of young children ever be a wise decision, in light of the impact of those two disciplines on brain function, self-worth, and a sense of community through the arts, for one thing? What was "pure" about it? And how could it not be personal, especially for Lucy, who had carried on the tradition established by her father? Not *personal*?

Three bullet points. Extra emphasis on the third.

All three stayed buried under seasoned meat and shredded cheese. Lucy sipped her raspberry lemonade for courage. "I understand." Wait? What? That's not what she'd planned to say.

"Well," Mrs. Schindler said, "best wishes in your future endeavors." She paused a long moment before leaving Lucy to her salad and super-sized lemonade.

Future endeavors. *I'm fifty-six. Almost. You chopped at least nine years off my future endeavors.* Not that Lucy seriously entertained the idea of retirement at sixty-five. Ninety sounded more reasonable, providing she could stay a little more alert than Kiersten's grandmother.

Her cell phone vibrated against the tabletop. She picked it up and checked caller ID. Charlie.

"Where are you, Lucy? Are you okay? Martin heard from Steve's mother that you were in a horrible accident. Why didn't you call?"

"Weren't you fishing?"

His pause made her think they'd lost the connection. "I still am. Reception's not great out here. Talk to me."

"Charlie, I'm fine. A small fender bender. In a parking lot. I already got an estimate from the body shop."

"You did?"

"It's crazy how little damage it takes to push a repair bill past the deductible."

"You *sound* okay."

Lucy had kept her voice low but now turned to face the corner of the wooden booth. "I am okay. I can't talk now, Charlie."

"Me either. Martin's got a bite, and it looks like he's going to need the net."

For a brief moment, she wondered if the two men were fishing with artificial lures or worms. "See you at home later."

Broken, but drivable.

6

He'd lied to her. He hadn't been fishing. He'd been painting the kitchen. From the looks of it, he'd started right after she left the house. If she hadn't stayed so long at Riverside Park, staring into the water and trying unsuccessfully to oust every verse of "The Water Is Wide" and "Down to the River to Pray" from her mind, she might have walked in on him mid-project.

Instead, she found him peeling the last of the painter's tape from the crown molding. He rolled it into a sticky ball and tossed it toward the plastic garbage can he'd positioned near the ladder. "Nothing but net," he said.

"Charlie, what are you doing?" She adopted her reserve-judgment-until-I've-heard-the-whole-story voice she'd often used on her students. And the Tuttle kids.

"Welcome to your freshly painted kitchen, LucyMyLight," he said, descending the ladder. "How do you like it? Baby, don't cry." He held her against his paint-speckled once-red shirt. "It wasn't that big of a deal. I can go fishing some other time."

She pulled back to catch her breath.

"Oh, look at that," he said, whipping a paint rag from his back pocket and daubing at the spot of Berrington Blue on her jean jacket. "The price of a hug, I guess. You'll want to wash that out right away, before it dries."

As if she hadn't done every single load of laundry in their house for the last thirty-two years. As if she didn't realize dried paint was a whole different animal than wet paint. As if she needed the instruction.

As if she needed him to take away the one project that would have made the summer have a small nugget of meaning to it.

"I had to hurry to finish before you got home. So we might have to do a few touch ups."

"It looks...nice, Charlie. I love the color." She wetted a corner of a clean rag and scrubbed at the paint smear.

"Me, too. You have great taste." He smiled. "Of course, we knew that already. You chose me."

He busied himself putting away the step stool, ladder, and his painting tools while Lucy shrugged out of her jacket to get a better look at the spot and contemplated how she'd ever pay for the counseling she probably needed. What kind of woman resents a man like Charlie? What's wrong with a woman like that?

Resent? Had she really used that word? The indictments against her character mounted like the constellations of Berrington Blue spatters on her white cabinets.

She worked late into the night putting the kitchen back to rights. If only blemishes on her soul flicked off as easily as the dry dots of blue paint responded to her thumbnail. Lucy replaced half the items she normally kept on the counters, along the backsplash. As infrequently as she used truffle oil, it didn't need to occupy space on the counter, no matter how artsy the bottle.

After supper, Charlie had watched a WWII POW movie while she worked, then kissed her goodnight and headed to bed early, smelling of the pungent rub he used on aching muscles. The whole bedroom would smell like that. Another thing she should be used to by now. Not just used to. Grateful for. His muscles ached because he tried to do something nice for her.

She found another dot of paint on the granite countertop. Flick. Gone.

A wave of satisfaction worked its way to shore from far out to sea. She felt it nearing, but so much debris had washed up on the beach during the previous days' tides, the wave diminished in intensity by the time it reached her. The room looked fresher than it had in a long time.

Charlie's retirement health coverage meant the impact of her job loss threatened their savings plan more than their daily budget. Olivia and SamWise had weathered the unstable years and emerged as adult versions of the joy of their lives. Nobody was in the hospital, rehab, or jail. Not every family could say as much. She'd had nineteen years at a job she loved, pursuing a passion with an endless, pulsing rhythm. Her husband—who sometimes impersonated Captain Oblivious—loved her and showed it. What was wrong with her? Was she auditioning for the most ungrateful human on the planet?

For a second, a split second, she understood why a woman with leftover pain medication might take something to quiet the internal condemnation.

Instead—and because she had no leftovers—she turned out the kitchen light and went to bed.

"Lucy? Lucy."

"What?"

"How long are you planning to sleep in?"

She drew the comforter over her shoulders. They could probably set the air conditioner temperature a little warmer and save electricity. "What time is it?"

"Eleven..."

"Eleven!" She threw the covers off with a snap like a mainsail in a stiff wind. "Why did you let me sleep so long?" Lucy sat on the edge of the bed, fighting to get her bearings.

"Eleven minutes after eight."

"Charlie!" Lucy fell back into her nest of pillows.

"The muffler's done. Can you drop me off at the shop so I can pick it up?"

"Now?" She scrubbed her hand through her hair.

He pulled off the work shirt he'd been wearing, sniffed it, then threw it into the hamper. "Did you have something else you needed to do?"

"Sleep?"

"Hon, you can do that anytime now that you're..."

A song from the animated movie *Frozen* flashed through her mind. Couldn't he let this go?

"Sorry I woke you, Lucy." His eyebrows scrunched forward. "You used to be up by six."

"I stayed up later than usual last night."

"Oh." He tugged a polo shirt over his head. "I guess I could call Martin or somebody to take me down there."

"No, I'll get up," she said. It was the least she could do. "Aren't you a little overdressed for the muffler place?"

"Once I get the car, I'm heading over to Silver Lake. A guy there has done some worm farming in the past. I'm going to pick his brain. Want to come?"

God help her, she'd reverted to her ugly self, and it wasn't even eighty-thirty in the morning. All she could think about was how slim the pickings would be.

"Want to come along? That would be great. We can talk the whole way there and b—"

She closed the bathroom door and started the shower. "Sorry. Can't hear you. Be out soon."

How many prayers had God heard over the years? How many of them came from shower stalls? And how many were wordless like hers?

She toweled off, scrubbed several layers of enamel from her teeth, scrunched her hair, and slipped into the bedroom without letting too much of the humidity escape into the room. Charlie was elsewhere in the house, humming loudly enough for her to hear. Like an eight-year-old prepping for his best day ever.

She dressed quickly, fairly certain an ex-worm farmer was unlikely to be a fashion critic. *Did that session in the shower mean nothing to you, Lucy? You're profiling now?* She formed an apology with no destination.

It seemed rude to humanity for her to go out of the house without under-eye concealer. So she took time for that and a minimum of other makeup before heading for the kitchen. Today, she would be grateful, patient, and optimistic. Grateful, patient, optimistic. She rehearsed all the way down the hall.

"Baked grapefruit okay with you?" Charlie asked. One of his few culinary specialties. She was... grateful. Genuinely. Maybe this was the day the darkness would lift.

"I'll make coffee."

"Done already."

Usually, coffee warmed her insides after she drank it. Charlie's thoughtfulness started the wave of warmth before she brought the mug to her lips. "Thanks, honey."

"You're welcome. It's the least I can do for my worm partner."

To speak or not to speak? Did every marriage wrestle with that question 24/7? "About that, Charlie." She sipped the coffee, mind racing, opening one door after another in her search for a suitable response. Doors were still banging shut when her husband slid a bowl with a caramelized grapefruit-half across the breakfast bar toward her. She stopped it before it slid over the edge. Could she stop herself soon enough?

"Oh, we'll find a better name for you than worm partner," Charlie said. "Executive Director of Wormology? Worm Princess? Secretary of Squirm?"

"Charlie!"

"What happened to your sense of humor, Luce? You should see the look on your face."

She didn't have to see it. She felt it. "I know you think you've found your life's passion."

"Which, I might add, you've suggested I needed for most of our married life." He guzzled his coffee as if proving he could.

Lucy practiced her lung-filling and lung-emptying breathing warm-up. "You haven't even talked to anybody about how that could work. *If* it could work. You don't know if you'd enjoy raising worms. Or what the market's like. Or how much it costs to get started."

"And that's why we're going to meet that vermiculturist guy today after we pick up the Traverse."

Vermiculture? That's what it's called? "Charlie, it's the 'we' that's a problem for me."

He set his coffee mug on the granite, folded his arms across his chest, and made a thin, lipless line where his mouth should be. The line softened. His arms dropped to his sides. "I don't know if you've noticed," he said, snatching his cap from the hook near the door, "but the 'we' has never been a problem for me."

"Where are you going?"

"I'll walk to the muffler shop. I'm sure you have better things to do."

Not to speak. That was the correct answer. And yet . . .

"Charlie, I don't mind going with you to the worm expert."

"Sounded as if you were *thrilled* at the prospect. I must have misread you." Sarcasm seemed so much harsher from Charlie than it did from anyone else. He removed his cap but didn't put it back on the hook.

She lowered her voice. "This is your passion. Not mine. But you're assuming we'll do this together."

"Would that be so horrible?"

Lucy considered herself a decent communicator in every area except this. She pressed her fingertips to her lips. The action didn't help her think any more clearly. "I love being with you."

"And that would be evidenced by . . . ?"

How could she blame him for what lay behind that open-ended question? This isn't the wife she wanted to be. Not who she was... deep down. Little of that fought its way to the surface past the oil spill of disillusionment.

"I get it," he said.

You get it that I don't know how to do this, how to take a breath past these cramped vocal cords, how to reconcile the fact that my husband is ecstatic because I have nothing left to do? Nothing left? And that I think that I may be slipping into an ugliness I won't be able to crawl out of... and I can't tell the man I've committed to love forever because it'll look as if I don't love him?

"What do you *get*, Charlie?" Her voice broke. She prayed he'd realize the cause wasn't disappointment in him, but in the turn life had taken. The hairpin, narrow, cliff-edge, crumbling, nauseating turn.

His gaze focused over her head. Maybe he, too, saw the dollop of Berrington Blue on the crown molding. "I get it"—he dropped his gaze to her eyes—"that I never should have changed deodorants. I'd be more pleasant to be around." He chuckled. "Come on, LucyMyLight. You have to admit that was funny."

His comedy act proved he really, truly, most sincerely did not get it at all.

She swallowed. The simple, no-thought-involved act didn't go well. "What time is your appointment with the Worm Whisperer?"

"Ten-thirty."

Lucy put her grapefruit bowl, untouched, in the sink. "We'd better get moving then."

"Look, don't come along if you don't want to."

"And miss the opportunity to expand my knowledge of a worm's digestive process?" *Courage, Lucy. Courage. It's only part of a morning. And it'll bless him.*

"That's my girl."

Our daughter Olivia is your girl. I'm your wife.

7

"Worm-casting tea?"

"Not to drink, Lucy. For fertilizer."

He was excited about fertilizer. Passionate about it. She cinched her seat belt for the drive home. "So, worms thrive on dead things."

Charlie glanced her way as he pulled the car onto the main highway. "You were listening?"

"Took notes," she said, holding her phone toward him.

"Sure you did."

He was skeptical? She opened the notes app. "'If it was once living and is now dead, worms will eat it.' 'Fifteen dollars for eight pounds of worm feed.' 'It can take as little as 350-400 square feet of space to produce almost eight tons of worms and more than two tons of worm castings per month.' 'A garage, basement, or even a spare room can be the scene of a starter worm farm since...'"

"Go on."

"'Since there are no offensive odors related to worm farming.'"

Charlie smiled, the lines from his nose to his mouth triplicating. "That"—he slapped the steering wheel in victory—"was great news."

A vermiculture headache clamped down hard on the top of Lucy's skull. "You believed that line?"

"That it doesn't take much square footage to get started?"

"About the odorlessness of this process? Dead things. They eat dead things."

"How many square feet is the guest room?"

"No. No, no, no, no. Nope. No."

Charlie's fingers played a nameless tune on the steering wheel. "Think about it a minute, Lucy."

"My head's about to explode."

"You're probably dehydrated. We'll stop at the drive-thru for something to drink."

She slipped on her sunglasses and leaned her head against the headrest. "Lemonade. I need more lemonade. Apparently."

"We should probably wait to talk until your headache's better."

"Good idea."

The blessed silence lasted less than a minute. Charlie's favorite radio station was recognizable by its unending stream of news. He could listen to the same series of news reports with no apparent annoyance at the repetitiveness, the mind-numbing, soul-deflating repetitive account of all that's wrong and twisted and broken in the world.

Her phone's earbuds lay in a zippered pocket in her purse. The thought of them tugged at her. Music waited on her playlist. Soothing, uplifting, intriguing, soul-fortifying music. Worth listening to. Worth the time and attention. Songs she could listen to again and again without—

Without caring about the repetition.

As if viewing it from the sunroof, she pictured Charlie in his corner of the front seats, content, and her in the far corner, earbuds firmly planted, content. But alone. A wall of differences making the two feet between feel like an ocean's width.

Not what she wanted.

After this many years of marriage, was it unthinkable they could each have what they wanted without trampling the other's interests?

"Did you hear that?" Charlie said.

What fascinating piece of news had she missed? "Hear what?"

"That little trivia bit they just did." He nodded toward the radio and turned the volume down.

"No."

"I would have thought— Never mind. Anyway, they said Native Americans used to tell how sick people were by how long it had been since they sang."

He let the words hover, not tagging any editorial or husbandly comment to their end. He glanced at her, then briefly at the glove compartment or something near it before returning his focus to the stretch of highway in front of them. Was he looking at hash marks on the road or farther—into their future?

She'd weathered crises before. What was different about this time? Why did it seem this trauma had no note of hope? Even when her parents died, she found solace in song. "Sometimes I Feel Like a Motherless Child" brought expression to her grief. She sang it in the garden and hummed it in the kitchen until it no longer stung but soothed.

"It's been a while," Charlie ventured, "since I caught you singing."

"I know."

"Or playing piano, guitar, anything."

"You're right."

Charlie swung into line for the drive-thru. "You haven't even made an attempt to line up private lessons or apply for a teaching position elsewhere, if it means that much to you."

If?

"Are you going to be okay?"

The word *eventually* made it as far as the tip of her tongue. Could she promise *eventually*? Lucy's cell phone vibrated. She wouldn't have to decide right now.

"Hey, Olivia. What's up? We don't usually hear from you during the day." A car length away from the order speaker now, Lucy said, "Olivia, can you hold for just a sec? Dad needs my order." She underscored that all she wanted was a tall lemonade.

Charlie nodded and placed their orders.

"So, what's up?"

"Mom, I made my decision."

Lucy adjusted the sun visor. "Go ahead."

"I'm going to work part-time this summer, rather than full-time."

"Oh?"

"So I can fit in some fast-track classes and start on my masters degree program."

If Charlie hadn't been paying for their drinks and ferrying them to the cup holders, she would have put the call on speakerphone right away. When he pulled the car away from the drive-thru window, Lucy said, "Olivia, Dad and I are both here. I'll put you on speaker. Okay?"

"Fine. Hi, Daddy."

"Hey, pumpkin. From the look on your mother's face, I'd say this is interesting news."

"You might see more of me."

"That's great," he said.

Lucy sipped her drink. Maybe they'd catch a glimpse of Olivia's mystery date.

"I'm only going to work part-time this summer so I can start working toward my masters."

Charlie looked Lucy's way. She nodded. "Sounds good to us, Olivia. We can't help financially, but..."

"No. I know that. I hoped you wouldn't mind if I crash at home more often, since I can save money if I sublet my apartment for the summer?" She ended with a vocal question mark.

Lucy held the phone aloft while sliding the straw she'd one-handedly unsheathed into her lemonade. "We'd love seeing you more often."

"I'd be in and out. It depends on where I can find part-time work that pays decent."

The eternal dilemma.

"I know God's got this," Olivia said.

Yes. That's what Lucy should have said. *God's got this.*

"But right now, I feel kind of lost."

Know the feeling.

"Three years out of college, and I'm still eating Ramen noodles for too many meals. But I have to go for it."

"We're always here for you," Lucy said. "No matter what's going on."

Olivia's sigh traveled through the phone and filled the car. "I know it will cost me money I don't exactly have right now."

"Join the club," Charlie said. "Just don't take the job your mother needs."

I do?

"Mom, you're getting a job?"

I am?

"Hey," Charlie said, his voice brighter than ever, stingingly bright, "maybe you two can find something at the same place." He leaned closer to the phone and added, "Your mom isn't signing on to my worm business, I guess."

"You're not serious, Daddy. You're going through with that?"

Finally. Reinforcements.

"Without a partner? Dad, come on. That could be a lot of effort. And you've never run a business."

Charlie's "I thought I had a partner" barely registered as audible.

8

They talked more details with their daughter until Charlie dropped Lucy where they'd left Lucy's defaced Malibu. She wondered if the term *vermiculture* originated in the similarities between vermin and worms, or if the "verm-" part was a Sergeant Schultz pronunciation of *worm*. Sergeant Schultz. *Hogan's Heroes*. TVLand. The places her mind drifted these days...

It wasn't until she'd strapped herself into her battered car that she realized Olivia's temporary homecoming would make what they called the guest room off-limits for worm breeding. Things were looking up.

When they got home, Lucy immediately aimed for the guest room. What could stay? What would have to relocate to make room for their grown daughter and her accumulation of things? When she'd stayed only a night here and there, the guest room had sufficed the way it was, with just a sliver of space in the closet. Lucy started a mental list.

"So..." Charlie entered the room behind her.

"Doesn't seem that long ago we converted this room from her college-break hovel to this much tidier version of a bedroom."

"So..."

"What is it?"

Charlie ran his hand along the edge of the chest-high set of drawers. "I made breakfast."

The grapefruit half she hadn't eaten. "Yes?"

"It's lunch time."

She wasn't hungry. He was. A good wife would—would what?

Her mother would have apologized for not realizing it was past noon, dropped everything, and raced through getting the breadwinner something to eat. Ania would have said, "I'll take a grilled ham and cheese. Thanks." Scratch that. A grilled eggplant and roasted red pepper. Lucy's sister—twice divorced—would have planted her hands on her hips and stared him down until he said, "I'll... I'll just go make myself something." To which her sister would have replied, "You do that," and muttered merciless adjectives under her breath. Olivia might say, without missing a beat, "Oh, you're right. Time flies. I'm not hungry yet, and I'd like to keep working here. If you have trouble finding something in the kitchen, let me know."

Lucy had been all of the above at various stages of their marriage and to varying degrees. She only planted her hands on her hips and gave the stare-down in her mind, never externally. And she wasn't that fond of eggplant.

The answer she most admired? Olivia's.

"I'll get you something in a minute, Charlie."

An answer like that was supposed to make her feel good about doing the selfless thing, like Jesus taught. Why did it make her feel as if she'd missed His larger point?

She tried singing while she made tuna salad. After the first few notes of "who lives in a pineapple under the sea?" she abandoned the musical soundtrack for lunch prep. Some of her students' influences weren't worth archiving.

"Want to eat on the deck?" Charlie asked when she called him away from the noon news. "Beautiful day out there."

During summer school, in an ordinary summer, she would have grabbed a nectarine and a bag of microwave popcorn for lunch between her packed schedule of students' individual lessons or the group lessons for beginning musicians. In nineteen years, she hadn't yet, but had intended to start taking the stairs or walking the track outside the school for exercise. This was the summer, she'd told herself. Best laid plans.

She looked at the dome of tuna salad on her plate surrounded by a halo of barbecue-flavored potato chips. She replaced one chip with a baby carrot. Good enough for today. "The deck. Sure. That sounds great."

It did. And it was. Enough of a breeze to keep the bugs away. Everywhere she looked, something was blooming, chirping, or flying. Under the lawn a few inches, grassroots earthworms created fertilizer for the green blades that fought dandelions for dominance.

Charlie opened the tabletop patio umbrella to shade the two of them from the high sun the maple tree failed to block. Lucy sank into her favorite glider patio chair and pulled up to the table.

Charlie prayed briefly over their meal, with an extra line thrown in for Olivia. Lucy wondered if Olivia had called her brother. She and Sam had become such good friends in adulthood. Who would have thought? They'd tolerated each other as toddlers two years apart. They'd loathed each other in high school, tormenting each other both intentionally and unintentionally. How had they moved from tolerating and tormenting, she wondered—watching Charlie smear tuna salad onto a potato chip—to admiration and respect?

They'd grown up.

Now, that was a dangerous thought.

"This is my definition of perfect," Charlie said, leaning back in his matching patio glider rocker, then leaning forward to grab a napkin to catch the dollop of tuna that landed at breastbone level.

"Perfect? The breeze helps. And the umbrella. Sometimes I wonder," Lucy said, not for the first time, "if we shouldn't have screened in this deck when we moved here."

"I wasn't talking about the weather."

Oh.

"This 'us.' Just us. Nothing to do but sit here and enjoy one another."

Her left eye twitched. She turned her head so he wouldn't see if it happened again. It did. Lucy couldn't imagine a world in which "nothing to do" and "perfect" fit in the same paragraph. A pouch of tuna, mayo, celery, potato chips, nothing to do, plus *her* spelled contentment for him. Her list didn't start with tuna.

His was the face she wanted to see at the end of the day, the eyes she most wanted to dream about, her favorite embrace. She craved holding hands all day every day when they were in college and never being apart far enough to slip a dollar bill between them, like testing for gasket leaks in a freezer door. At this stage of life, the idea made her claustrophobic. Guilt clamped iron fingers around her upper arm as if hauling her to relationship jail. He wasn't her teenaged, annoying, flatulence-obsessed older brother. Except for the flatulence and older. He was her husband. Hus. Band. For life.

Other summers, he'd worked and then come home at the end of the day. Other months of the year until this past year, they'd both worked and come back together at the end of the day. Lucy separated the celery from her tuna salad with her fork.

Something landed on her forearm. She brushed it away. His hand. "Sorry. I thought you were a June bug." Perspiring more than necessary for a day with a soft breeze, she reached for his hand and replaced it on her arm.

Charlie's face broke into the irrepressible grin that had fluttered her heart so many days of their marriage. "A June bug. That's a compliment, right? Because I know how much you l-o-v-e June bugs."

"You startled me. My mind was elsewhere, I guess." Actually, it was right where they sat, but try explaining that.

His fingers traced her arm from fingertips to shoulder and back again. And again. "Lucy, you know I'm on your side, don't you?"

"In what?"

"Battling this depression."

"I'm not depressed." *I'm sad. That's all. Incredibly, deeply, soul-woundingly sad.*

"Sometimes . . . when a woman reaches your age . . ."

She bit the side of her tongue. "Menopause? You think this is from menopause?" She'd check a mirror later for broken blood vessels in her eye.

He stopped stroking her arm and wiped the condensation from his glass. "Maybe it will be good in more ways than one for Olivia to move back home for a while."

The great shrinking house. Did he think he'd have reinforcements when Olivia got there? "It'll be good to have her here. For a while," she agreed.

Lucy needed to turn a room for overnight guests into a longer term guest room. She had flowerbeds that needed water and weeding, not in that order, and she craved a few minutes on the computer to look up clinical depression so she could prove that wasn't her problem. Charlie might have all day, but she had things to do.

"It's funny Martin hasn't called lately to take you fishing with him," she said.

Charlie swallowed, most of his mouthful, and said, "He calls almost every day."

"He does?"

"I turn him down."

Lucy's eye twitched again. She pressed a finger against the offending muscle. "Why would you do that?"

"You've been kind of needy lately." The tilt of his head and almost impish look—so like Sam's—said he didn't mean it like it sounded.

She forced her face into neutral rather than the glare that begged to form. *What I've needed is a little breathing room.* She should tell him exactly that. Yes. "I think you should take him up on it one of these days. Go. You need to get out on the water while the fishing's at its peak before it gets hotter and the fish aren't eating as voraciously as they are now."

"Listen to my outdoorswoman here."

"I've heard your fishing stories for a good number of years, Charlie. And, for the record, I'm not needy." She flicked her celery bits back into the pile of tuna and stirred extra long.

"Did you get Martin on the phone?" Lucy grabbed another armful of out-of-season clothes hanging in the guest room closet.

Charlie leaned against the doorjamb of the room that would become Olivia's again. "Yeah."

"And?"

"We're not going today."

"Okay." She shifted the weight draped across her arm. "Why not? Too late in the day?"

"Too late"—he sighed—"in life." He lifted the bulk from her arms, laid the clothing on the empty bed, and swallowed her in his fierce embrace.

"Charlie? What's...?" She could feel him trembling.

"Eve had an appointment today," he breathe-spoke into her hair. "Her doctor thinks it's early onset Alz—"

"No!" Lucy matched the grip of his embrace. "Martin must be devastated. Oh, Charlie!"

"Don't leave me, Lucy."

"I won't."

"I mean, mentally. Or...emotionally. I don't even know what I'm saying. Just don't. Okay?" He rocked her back and forth. Almost a dance. Not quite.

"What are you whispering?"

His dance slowed. Stopped. "Praying for Martin."

Lucy pulled back. She wiped the moisture from the corner of his eyes with her thumb, and marveled that two polar opposite hearts could beat as one. "And I was praying for Eve."

She smoothed the curls of silver in front of his ears. The man needed a haircut again. In her head, the melodies of every Top Forty love song tumbled over each other. Yes. This man. Forever. Somehow.

9

Charlie turned away from her a few degrees. "I shouldn't have retired early. Shouldn't have done it."

"At the time, we had no idea I'd be riffed." Lucy picked up the stack of clothes from the guest room bed. He followed her to the hall closet.

"If I'd held out until sixty-five, I would have gotten my full pension. That would sure help right now."

"Charlie..."

"No, let me say it. We both know it's true."

Lucy nudged him with her foot. "The closet, Charlie."

"What?"

"Would you please slide those coats to the side so I can fit these hangers on the rod?" Should have been obvious, shouldn't it?

"Oh. Sure."

Now was not the time to remind him she'd been against the idea of early retirement. "We took a calculated risk."

"It all made sense on paper. Before..."

It never made sense to her. She couldn't imagine longing to retire from teaching music. That was so secondary to the real crisis. Any town whose school board could extinguish the music and art programs—

"I'm not blaming you." Charlie grabbed one of his wrists with the other hand and rested them on top of his head, making a diamond out of his elbow angles and as a sidelight reminding Lucy he'd asked her to pick up deodorant the next time she went shopping.

Not blaming me? Good. Since it wasn't *my fault. Choose your words more carefully, sir.*

"I mean, it's not your fault, of course."

"Correct." How did they get the one Craftsman house built in the '90s without wide hallways? Lucy shimmied past him to get another armload from Olivia's room.

"It will be a while before my business gets off the ground."

"Uh huh." *Hi, I'm Lucy Tuttle. My husband makes worms.* "A thing like that takes a while." *God, I apologize for what may have sounded like a small, miniscule really, tidbit of disrespect. Lord, even You must find worm farming a little ridiculous, right?* Where was this conversation headed?

"I can't see trying to hold down another job during the start-up. I've got my business model to set up. Construction. Purchasing. Marketing."

She waited for him to continue.

"And..."

"What is it?"

"There will be...start-up costs." He sounded matter-of-fact and hesitant at the same time. How did he manage that? "And with Olivia moving in now..."

"Are you rethinking this?" She sank onto the edge of the twin bed, a glimmer of hope in the bounce.

He sat beside her. "Well, yes."

God answers prayer. He does.

Charlie took her hand in his. "LucyMyLight, I don't see any other option."

No worm bedding in the Tuttle house. Lucy remembered what giddy felt like.

"No other option than for you to get a job."

A revolt of muscles and nerve endings started at the base of her neck and inched upward. "That's your solution?" She gets a job to pay for his hobby's start-up costs.

He stood then and paced. "Not ideal, of course."

"No. Not ideal." If he hadn't been pacing, she would have.

"Not at all what I'd hoped for," Charlie said.

"Which was?"

Charlie stopped pacing. "Us. You and me. Working the vermiculture together. Side by side. Never having to be apart."

Was he pleading his case? Did he think that all sounded appealing? All of it? "You assumed that's what I'd do with the rest of my life."

"No. Of course not."

Hope for him yet?

"We won't be worm farming when we're in our eighties, I don't think."

Every muscle tensed. Her entire nervous system misfired. Maybe she had fibromyalgia. Or that other one. Chronic fatigue syndrome.

No. She had Charlie. For life.

When Ania the Angry Artist called to invite her for lunch the next day, Lucy almost said no. In a bizarre twist of total lack of logic, Lucy thought she might be able to make a difference in the young teacher's life, might influence her for good.

As long as Ania asked no relationship advice. Or questions about the role of faith in hopeless situations.

They drove to Woodbridge, which had a better selection of restaurants and offered less possibility of being overheard by someone they knew.

"Where do you want to eat?" Ania asked, pulling her duct-taped Gremlin-like foreign car into the mall complex. Lucy had volunteered to drive, but Ania felt it would be tacky to be seen in a Malibu with a bite mark in its bumper. Ironic now. Ania won the argument, despite the duct tape. No one could deny that's exactly what the injury to Lucy's car looked like. A bite mark.

"We've never eaten at that new sushi place," Lucy said.

"Mostly because I don't care for sushi."

Lucy craned her neck to catch Ania's facial expression. "Really? I would have thought you'd love sushi."

"Because I'm all artsy and trendy?"

"Or..."

"Because I'm Asian?"

"Yeah. That."

"I'm also eating vegetarian."

"There are a lot of veggie sushi options, aren't there? Or have it made to order with just rice and...cucumbers...and..."

Ania drove with her wrists on top of the steering wheel, her arms stiff, a habit Lucy had seen her revert to when thinking deep thoughts.

Lucy scanned the wealth of restaurants in the vicinity. "Panera?"

"Ate there yesterday."

"You choose then."

Ania gripped the wheel at nine and three. "There we go. Made you say it."

"What?"

"My choice?"

What was Lucy missing? What evil did Ania have planned?

The young woman—wearing her signature frayed jeans, tank top, and infinity scarf plus gallery-worthy jewelry—stayed tight-lipped until they'd navigated most of the frontage road and its appendages. She brought the car to a halt in front of an obscure stretch of storefronts.

Lucy tried to imagine which of the suspicious-looking buildings was Ania's destination, and whether any of them were involved in illegal activities.

"Come on," Ania said. "This will be fun." She exited the car like a woman less than half Lucy's age. Which she was. She stood at the entrance of an establishment with the windows blacked out. If she'd had a piece of bread in her purse, Lucy would have dropped crumbs to mark where she'd been before she disappeared, should the authorities need clues.

Lucy's eyes needed no time to adjust to the interior lighting. "Reminds me of an art gallery," she said to her companion, in a subdued voice befitting the setting.

"It is."

"And a restaurant."

"Which is only one thing shy of what makes it perfect."

"One thing?" Lucy took in the artwork on the walls, a combination of watercolors, oils, and photographs, lit to reflect the quality of the work, the unique table arrangements and place settings, the pale birch wood floors—spotless—and the cadre of wait staff in all white.

"There," Ania said, pointing to an alcove to their right.

As if cued by her voice, a cellist pulled his bow across grateful strings. Art, music, the promise of great food. Perfect. "How did you find this place?"

"I know the owner. Friend of mine from college. And very possibly my next love interest."

Oh, no. Relationship questions on the horizon. "Let's not get our paintbrushes before our palettes, Ania."

"Or our batons before the downbeat?" Ania approached the hostess stand and asked for a table for two near the Impressionists.

Lucy wouldn't have minded a lack of conversation in that atmosphere. She breathed a little more deeply than she had for a few days. Weeks. She clasped her hands together, rested her elbows on the unique zebrawood table to which they'd been led, and soaked in the nuances of the environment.

Something soul-refreshing in every corner, including the air itself. Wildflowers in pottery vases on the tables and divine aromas in the air.

"That smells like seared meat, Ania."

"Oh, I hope so."

Lucy unrolled her napkin and noted the antique silverware around which it had been wrapped. "You hope so? I thought you were eating vegetarian."

"With one exception. The Gallery's avocado burger with morels and crispy shallots."

"You purist, you."

Ania pouted. "It has lettuce."

Lucy snickered.

"And Lafferty. It has Lafferty."

"Laughter?"

Ania nodded toward the man approaching their table.

"Ania, great to see you here again." He kissed her on the cheek and extended his hand toward Lucy.

"I'm Lucy Tuttle, Ania's friend. We teach...we taught at the same school."

"Ah. The miffed," he said.

"Riffed," Lucy corrected.

Ania's sparkling eyes told Lucy her friend enjoyed his faux pas.

"Reduction in force." Instantly sorry she'd pursued the issue further, Lucy smoothed the napkin in her lap and said, "Love your restaurant."

He lifted his gaze ceiling-ward. Was he carried away by the cello's sonorous notes? Or calculating how much he'd invested? At length, he spoke. "Dream come true." He drew in a deep, satisfied breath. "It's an amazing thing when your passion and calling and job intertwine."

The now familiar ache spread outward from Lucy's core. No explanation needed. She knew exactly what he meant.

"Speaking of which," he said, directing his attention back to Ania, "have you made a decision yet?"

Wait a minute. How far had this relationship gotten? Wouldn't Ania have told her about Lafferty before now if it were serious enough for her to have decisions to make?

Ania turned when Lucy tried to catch her gaze. "Yes. The answer's yes."

Hold the fermata! Lucy should have started talking relationship wisdom the minute she climbed into Ania's faux Gremlin.

"Yes," she said again, still avoiding eye contact with Lucy.

"Great." Lafferty seemed only moderately pleased. "Monday, then? We're closed, but that's the best time for orientation. Thanks, Ania. We've been lost since our dishwasher quit on us."

Ania snuck a glance at Lucy. "And...then I'll have opportunities for other positions as they open?"

"Oh, sure," Ania's new love interest said. "Might not be for a while. You're okay with that?"

She looked directly at Lucy this time when answering. "I'm grateful for the job. And for the chance to be part of what you're doing here." She pointed toward the cellist and the paintings, her gestures an awkward imitation of Vanna White.

When Lafferty headed to other responsibilities, Ania said, "Please don't give me the 'this is beneath you' speech."

"Why would I do that? None of us is *beneath* any meaningful work, are we?" Oh, stab. Like worm entrepreneur.

"I'll be surrounded by art," Ania said. "I may even be able to get some of my own pieces added to the gallery. Wouldn't that be great?"

"It would." Lunch. She wanted lunch with her friend. Maybe some safe conversation. Time away from her shrinking house. This was not on the agenda. "If that's how you want to spend your summer, go for it."

Ania folded her napkin into accordion pleats. "Maybe not just the summer, Lucy."

"Oh?"

"I love art." She traced her finger along the contours of the pottery vase in the center of the table.

"And I love music," Lucy said. The cellist filled the spaces between molecules in the room with a tender rendition of the prelude from Bach's Cello Suite No. 1. She could imagine a thousand scenes it might illustrate.

"That's not the whole truth." Ania emphasized the word *whole*. "You're passionate about *teaching* music. And"—Ania's voice nearly disappeared—"I'm content to make art, to be around art. The teaching part?"

Lucy held her breath. "You're not going after another teaching position elsewhere?" Was Ania's heart beating as oddly as Lucy's? "Don't sell yourself short. You're a great teacher."

"I can do it."

"Wonderfully, by the way."

"I couldn't understand why you were devastated instead of angry. I finally reached devastated. My first year as a teacher and it ends like this? Maybe I'm supposed to walk away. You can't imagine how you could."

The air changed. The waitress at their table brought the fragrance of self-tanner with her.

"Can I take your drink order?"

It wasn't just the sweet, beachy smell of tanning lotion. The music stopped. Lucy glanced at the alcove. The cellist had laid aside his cello and bow. He stood and walked toward the hostess stand. *Don't leave. Please don't leave.*

"What do you want to drink, Lucy?"

She drew her attention back to her friend and the stranger ready to fill their drink orders. "Lemon—no. I'll have an iced tea. Unsweet."

10

The burgers were as good as Ania claimed. Good enough to temporarily convert a vegetarian.

Ania's revelation had sobered Lucy. Ania had always treated teaching as more than a nine-to-five job, a job she performed well. Lucy could pick out people who pursued teaching as a career versus a calling, and those who taught because they knew their subject but didn't know children or how to reach them. Ania wasn't one of them. Did some of those teachers feel as trapped *in* their jobs as Lucy felt locked *out* of hers?

"You know what they say," Ania offered as the conversation wound down. "'If you can live without teaching, please do. Your students will thank you for it.'"

Where did that leave Lucy? She *didn't* think she could live without it. And no alternative appealed.

"It'll all work out for you, Lucy." Ania ate the last in the basket of homemade or rather handcrafted potato chips drizzled in truffle oil, sprinkled with shaved parmesan, the kind of treat Lucy knew would soon become mac and cheese if Charlie didn't stop dreaming up new business ventures.

"It'll work out for you, too, Ania. Don't give up on your teaching career, though, because of this one experience."

"I only have my own wits with which to figure it out. You have help."

"Charlie isn't giving me much insight about where to go from here. The one idea that appeals to him doesn't appeal to me."

Ania snatched the bill for the meal before Lucy realized what she'd done. "I wasn't talking about Charlie." She slipped her credit card into the black leather folder and handed it to the waitress. "I meant that God-connection you have."

Lucy had taught her students to keep counting during the blank spaces, the rests. She'd urged them to maintain their intensity even when producing no sound. Rests had dominated the musical score of her "God-connection" since the RIF letter. She'd groaned through the rests. Messy. Hollow rather than soundless. Ania assumed Lucy had an advantage because of her faith in a God who hears, listens, and responds to those in need. Did she?

Or when her plans fell apart, had she taken her eyes off the director's hands? In her classroom, that usually spelled instant train wreck.

Ania signed the credit card receipt while Lucy entertained a mental picture of God flailing His arms to get her attention. He was thinner than she'd imagined. Probably from all that exertion.

The two women hit the resale shop after lunch. Nothing caught Lucy's eye except a repurposed cupboard door, hinges removed. The door inset had been painted with chalkboard paint. In white, a craftsperson had written, "Some days there won't be a song in your heart. Sing anyway. —Emory Austin."

She stared at it while Ania searched the vintage clothing racks. When it was time to check out, Lucy left the saying where it was. How would it help her to haul it home and have its message nag her every day? If she were still teaching, she would have snapped it up and hung it in the music room.

Another stab. She could teach the lesson. Why couldn't she live it?

"You didn't find anything?" Ania said, laying a small stack of items on the checkout counter.

"Still looking," Lucy said. In more ways than one.

"You need more time in here? I can wait."

"Time isn't going to help."

"What were you interested in? Besides that chalkboard piece you couldn't put down. I know. Let me get that for you. It's a little early for your birthday, but—"

"No!"

The resale shop checkout lady's eyes widened to the size of the enormous buttons on one of the jackets Ania had picked.

"No," Lucy said more calmly. "Thanks. I found it...interesting...but I'm trying to...declutter."

"Oh." Ania took her time turning her attention back to her purchases.

"I'm repainting the bedroom one of these days. Going to go more...minimalist."

Ania shot her a quick glance. "I thought you'd hang it in your office."

"Office. Yeah. Charlie said that if I didn't need my home office anymore, maybe he could use that for his vermiculture start-up."

Ania turned her whole body toward Lucy this time. "The room across the hall from your *boudoir*?"

"We don't really call the master bedroom at our house a boudoir."

"You voiced your opinion about his taking your office, didn't you? Didn't you?"

Some days, my friend, it's hard to know the difference between patience and enabling. If you and Lafferty get serious, you might find out. "I told him we'd have to think through the ramifications."

"So three seconds later you two decided...?"

Lucy grabbed a packet of gum from the impulse rack and then returned it. Gum from a secondhand shop? Even new, that sounded wrong. "We haven't gotten back to the subject yet. He's still in research mode."

Ania shook her head, her chandelier earrings clattering with the movement. She stopped abruptly and grabbed at her ear. Her face twisted as she worked to untangle the earring that had caught in her gauzy infinity scarf. "A little help?"

Lucy took over, tracing the thread that had wrapped itself around the setting of one of the stones on the earring. "I don't know if I can untangle it while you're still wearing that light fixture."

"Very funny. Here." Ania removed the earring.

Loosed from what anchored it to Ania's head, the earring was easily released from the scarf trap. Ania dropped the earring in her purse, temporarily, and paid her bill. Lucy spent the time wondering how to unanchor the feeling of defeat stuck to her head.

Some days there won't be a song in your heart. Sing anyway.

The more she thought about it, the more she knew she'd made the right decision leaving that chalkboard art at the store. Ridiculous message. If not ridiculous, at least unrealistic.

The clerk stuck a flyer into Ania's bag of purchases and offered one to Lucy, too, the nonpaying customer. The flyer advertised their upcoming sidewalk sale.

Nice one. The artwork on the paper was a section of weathered garden gate—turquoise—with the words:

> *Don't save your song for the shower.*
> *Don't save your words for the car.*
> *Shine your light in the darkness.*
> *Speak life wherever you are.*
> —Kathy Carlton Willis

Retail therapy? Not all it was cracked up to be. Tomorrow she'd look up the number of the therapist her doctor had recommended weeks ago. *God, I could use some help, here.*

11

The approach to the house—the curved flowerbed that almost obscured the front door during the summer and fall—needed attention. Perennials fought valiantly against bullying weeds. If it cooled off later, Lucy would spend some time eradicating the troublemakers.

She turned and waved as Ania's duct-taped car pulled away from the curb. Something had shifted in Lucy. She found it almost comical that Charlie opened the door from the inside before she could reach for the doorknob.

"How was lunch with the rebel?"

"Good. Delicious. Charming. Enlightening."

Charlie handed her the day's mail. "Enlightening. It's not every day that lunch at Bernie's can be labeled enlightening."

"We drove to Woodbridge and ate at a new place near the mall. Great food." The mail included a note from Ellie's mom. She'd read it in a minute. "Ania's going to work there. She starts orientation Monday."

"What do you know? Good food, huh?"

She gave Charlie a pass for not digging deeper. "Delicious."

"Must have been. Lunch took three hours?"

How many passes did she have in her reservoir? "Lunch only took an hour and a half. Then we stopped at a secondhand store. The rest was drive time."

"Oh. Yeah."

"In other news," Lucy said, "God and I are thinking about getting back together." She let the sentence hang and made her way to the kitchen for something to drink. He followed. The refrigerator door felt sticky. She grabbed a paper towel, squirted it with waterless hand cleaner, wiped the handle clean, swiped at an unidentifiable spot on the water dispenser, and opened the fridge. "Are we out of lemonade?"

"I drank the last of it waiting for— Drank it earlier."

"That's okay. Water's better for me." She filled a glass with ice and water from the dispenser.

"I didn't know you two had broken up. You and God, I mean."

"Temporary separation. It was all a misunderstanding." She couldn't afford to do more than make light of it at the moment. She needed a quiet corner and...and...

"I have news, too." Charlie tucked his index fingers in the corners of his mouth and exaggerated his smile.

"You need a root canal?"

"Nope." He increased the upward curve.

"Charlie! You're growing a moustache!"

"No. I didn't shave this morning, I guess." He growled and rubbed the edge of a hand across his upper lip, wincing. "Something else."

It could be any number of a million things, knowing her husband. "You're going to have to tell me."

"My face doesn't look swollen to you?"

She squinted and stepped closer. "I don't know. Maybe a little around the eyes."

"You should have seen me two hours ago." He gestured with his hands spread wide, fisherman-like. "And you know that strange rash I sometimes get after fishing?" He extended the backs of his hands.

"I don't see anything unusual."

"Well, it was worse earlier." He rubbed at the fading redness.

"I didn't think you were going fishing today."

Song of Silence

"Didn't. See that box on the counter?"

The shipping box that took up half the space between the stove and sink? Yes. She'd noticed.

"My test kit."

She set the water glass on the opposite side of the sink and peeked into the box. A cube-shaped Styrofoam cooler.

"Go ahead," he said. "Take a look."

"Worms?"

"Only a gross of them for now. They were more expensive than I thought. Plus shipping."

Gross is right.

"Charlie, I didn't know we—you—were ready to invest the... We hadn't decided to spend the— Charlie!"

"Test kit. As I said." He kept his distance. From her or from the box? Curious.

"The kitchen counter? Really?"

He took another step back. "Worms are cleaner than the average toilet seat."

"As encouraging as you meant that statement to sound—" She picked up the box and headed for the deck. He scooted to the opposite side of the room. "Can you get the door for me?"

"Probably shouldn't."

She stopped, her arms full of boxed worms. "What's going on?"

"I think I'm allergic."

"To Styrofoam? Never heard of that." What kind of crazy idea would he come up with next? Allergic to— "You mean, the *worms*?" Shock. Awe. Possible answered prayer.

He scratched the back of one hand, then the other. "I called Dr. Brandt's office. Have an appointment with him tomorrow afternoon. He called back a few minutes ago. He looked it up. There aren't many documented cases, but it's possible I'm allergic to earthworms."

"No living way." *I feel a song coming on. Stifle, Lucy. Stifle.*

"This is no laughing matter."

"I'm not laughing, Charlie." Not on the outside. "Look, I'm all about listening to your story, but can we get these nasty beasts out of my kitchen?" He didn't inch closer but held his itchy hands up in surrender... or helplessness. She set the box on the floor, opened the door, held it open with her foot, retrieved the box, and stepped onto the deck.

"They can't get overheated," he called from deep inside the kitchen.

"What do you expect me to do with them?"

"Basement? Until we get this figured out?"

Into every woman's life comes a defining moment, a moment when time pauses long enough for said person of sound mind to rise to a challenging occasion with heroic bravery. This was Lucy's moment.

She kept walking.

"Where are you going?"

"The garden."

"LucyMyLight, no!"

Heroic Lucy briefly wondered if a gross of worms constituted an overload for a small backyard garden. Still, she persevered in her quest.

"You can't do that." Charlie's athletic shoes slapped noisily across the deck behind her, but from the sounds of it, he drew no nearer.

"Odorless, my foot," she said, turning her nose upwind.

"You haven't seen the bill yet."

She stopped.

"And," Charlie said, his voice a swollen bundle of defeat, "they're nonrefundable. But I might be able to sell them on eBay."

How could that be legal?

Into every woman's life comes a defining moment, a moment when time pauses long enough for said person of sound mind to do something outrageously and undeservedly kind for the man to whom she pledged her life, in good times and bad. She turned on her heel and headed for the basement.

If she'd had her quiet space in which to discuss making up with the Almighty and asking His forgiveness for neglecting their relationship too long, Lucy might have been more agile conversationally when Charlie greeted her return to the top of the basement stairs with two disparate sentences.

"Would you mind washing your hands, LucyMyLight?"

As if she'd inadvertently or intentionally driven her fingers deep into the squirming ball of allergen-producing slime.

Then...

"So, I guess we've *both* lost our passions, huh?" He said it with a soul-crushing sigh.

Was every marriage this schizophrenic? Warmth followed by subzero freezer chill? An act of sacrificial kindness followed by a desire to remove his nose hairs with pliers?

"I...love you, Charlie."

"You say that as if you had to force yourself." He chuckled then sobered when she didn't join him.

"I'm going for a walk."

"A walk will do us both good," he said.

"Alone. I'm going for a walk by myself."

"Oh. You and God."

Right. She kissed him on his mildly swollen forehead and slipped her cell phone into her capris pocket. "See you later."

"You'll be back in time for supper?"

He had put a question mark at the end of that sentence, hadn't he? Not a period.

"We might have to eat a little later than normal tonight, but yes. I'll be home in time to make supper."

"Not you."

"What?"

Charlie crossed to the fridge and pulled something out of the waist-level drawer. A package of hamburger. "I thought I'd grill some burgers for us. There's leftover deli potato salad, I think. That should do it."

She could still taste the crispy shallots and morels from the burger she'd had at noon. "Sounds... great. Thanks."

"So give me a call when you're getting close to home and I'll fire up the grill."

Close to home. *Some days that's harder than others.*

She changed clothes and slipped out the front door. Despite the humidity in the air, she could breathe easier as soon as her shoes hit the sidewalk. "God, I'm so messed up." She retied one shoe. "And this is not how I intended to start our conversation." She walked to the end of their block and turned right, in the direction of the hiking path that traced the shoreline of the river. No other words came. She focused on the act of walking. It took more concentration than it should have.

Her chest tightened. Apparently that's where unspent tears are stored.

Too many days had passed since she last walked along the river. Her calves chided her. She stopped at one of the stone benches that dotted the grassy easement on either side of the path at quarter mile intervals. With one foot on the seat of the bench, she leaned forward then backward, stretching muscles and ligaments. She did the same with the opposite foot, but stopped when she noticed the inscription carved into the backrest.

> *Time spent learning*
> *how to love better*
> *is never wasted.*

"That is not my problem. Survival is my problem. Knowing what to do next is my problem. Charlie is my—"

Pain curled her toes inside her right athletic shoe. She stomped her foot, rotated her ankle, shook it, tried walking out the cramp, and finally resigned herself to sitting on the finger-pointing bench until the—huh—*charley* horse passed.

One of the things she appreciated about this river was the way it frustrated city planners. Long ago, it had picked a path that wound through what would become the bedroom community that grew up around it. Easier on street planners, a straight channel wouldn't have offered the kind of vignette-like scenes the Willow River created with its wide-sweeping, double-back curves. Birches, hardwoods, willows, and pines hugged the river's margins, the perfect mix for year-round beauty. She pulled out her cell phone and took pictures while she stretched the cramped foot, then pocketed the phone again.

Learning how to love better. Never wasted.

The words seemed to embed themselves in her back, through the fabric of her exercise shirt. She bent forward slightly, rubbed her palms on the thighs of her workout capris, and said, "Okay, God. Time to get real."

Her phone rumbled, startling her upright. She glanced heavenward while retrieving it from the handy sports pocket. *I didn't expect You to call,* she said to herself, the tension broken by technology. The caller ID warmed her heart. "Olivia."

"Mom?"

"Daughter. Where are you?"

"That's why I'm calling. I'm here at the house. Where are you?"

Lucy switched the phone to her other ear. "We didn't expect you until Saturday."

"Yeah, well, change of plans. The person subletting my apartment needed in right away."

Lucy cringed. "That's not convenient for you, is it?" She pictured Olivia's small apartment crammed with partially filled moving boxes.

"I jammed everything I could into my storage unit in the basement and stuffed the rest of it into my car. So..."

Lucy's mind raced through all she'd yet intended to do to prepare for Olivia's homecoming. The room was ready, for the most part. She was going to pick flowers from the front of the house. Something to show her daughter was welcome.

"Where are you? Dad said you went for a walk. He seems a little lost around here. Like he's a loose paperclip rattling around in a box."

Charlie's a loose paperclip, and I'm a loose hinge. Don't we make a pair? "I'm heading home now."

"Want me to come pick you up?"

You know, God, I feel like a mother of toddlers who can't even go to the bathroom without tiny little toddler fingers sneaking under the door to be close to me. Kind of precious. Kind of suffocating. "I'm not far, Olivia. I'll be there in a few minutes. Need to work the kinks out of my underused calf muscles."

"Okay. Dad started the grill. I can't face unloading my car yet, so I'm making a tossed salad. Do you have any pine nuts?"

"I'd have to check. On my way."

"Want to talk me through where they might be while you walk?"

"Hon, if I could walk fast and talk at the same time, I wouldn't have needed the exercise. See you at home. In a few."

Never wasted.

The imprint of the letters felt like sunburn on her back.

12

This much stuff strewn around, and Olivia hadn't unloaded the car yet? First thing in the morning—family meeting.

Lucy stepped over a leopard-print duffle bag on her way to the kitchen. "I'm home."

"Great." Olivia laid aside a paring knife and gave her a monster hug. "I'm almost done here, but I need those pine nuts, if you have them."

Lucy pulled her shirt collar away from her neck. "Okay if I take a quick shower first?"

Olivia resumed chopping. "Dad's been done with the burgers for a while. Trying to keep them from drying out. Can you wait until after we eat?"

Just like toddler fingers wiggling under the bathroom door. Can't even take a shower on my own timetable. "I suppose. I'll sit downwind." She opened the spice cabinet, dug around among the items on the second shelf, and produced the longed-for pine nuts.

"They'll be better toasted," Olivia said. "Can you do that, quick?"

Lucy dumped a handful into a dry sauté pan and turned on the flame.

Olivia rinsed her knife and said, "Mom, you need better knives. These are beyond resharpening."

"No argument there. Christmas is coming."

Olivia chuckled. "Six months from now."

Lucy continued to swirl the pan to keep the nuts from burning. "As soon as you can—"

"I know, Olivia. As soon as I can smell them, they're done. Oh, I have to tell you about the great restaurant where Ania and I had lunch."

"I already know. Dad told me. Sounds like a great place. If either of us can afford it, we should go there sometime together. Let's pray we both find jobs."

"Not sure I'm ready for that yet."

"Going back to work? Finding another way to use your teaching skills? I figured you'd want to. You've never been particularly skilled at doing nothing."

Lucy loved the easy rapport she now had with Olivia. Having her around for a while could be a very good thing. "No, I haven't."

"Do you think there's any hope the school board will reverse its decision before the new school year?" She waved her father in from the deck.

Hammers and drills pounded in Lucy's head. She heard the sound of electric saws and could almost smell the piney sawdust in the room that had been reserved for music since the school was built the year before her dad was hired.

Her dad. What would her father have said if he'd been alive to witness this? The program he constructed note by note, the one he practically handed to Lucy when he died, gone without an echo. She hadn't been able to keep it alive until the school's financial picture changed. If it ever would. His masterpiece music program had been taken out of its frame and made into a tray for chips, dip, and red plastic cups of powdered drink mix.

"Hey, LucyMyLight. About time you came home. I was ready to send search-and-rescue after you." Charlie balanced a platter of grilled meat in one hand and punctuated his sentences with a giant pair of tongs in the other.

"Or you could have called," Lucy said, ignoring the implication she'd been gone longer than she deserved. "It's why I take my phone with me."

"Ready to eat, girls?"

Lucy caught the look Olivia flashed. *What?* They sat at the kitchen table. Still too smoky on the deck. Charlie removed his barbeque mitts, tossing them over his shoulder to land who knows where, and reached out a hand to each of the women.

"Are you sure?" Lucy asked. "It's been less than two hours since I touched earthworms."

Olivia looked from one parent to the other. Charlie leaned back to give his laughter room to grow. "Yes, I'm sure."

"Dad?"

Charlie took their hands and bowed his head.

"Dad? Is your face swollen?"

"Little bit. Lucy? Would you? Please? That song."

Since Sam and Olivia were infants, Lucy led them in singing grace over meals. Not the traditional "Be present at our table, Lord." A song she'd written. About gratitude and wonder and the joy of love around the table. She hadn't sung it since their nest emptied.

Olivia squeezed her hand. "We'll help, Mom."

Is this the farthest music would reach from here on? The circumference of this small table? A simple gratitude song? Was she done? Singing her benediction?

"I don't think I can."

Charlie nudged her with his elbow. "You have to try. You know you want to."

Shrapnel from the land mine Charlie stepped on flew in all directions. It must have pierced an artery. A person doesn't bleed out that fast without having nicked something important.

Lucy pushed away from the table. She stood, her focus on the musical note salt-and-pepper shakers in the middle of the table. "I...don't have to try. And I don't want...to try." She slid her chair to an unoccupied position. "I want to take a shower. And then I'm going to bed. See you in the morning."

She could only imagine the conversation transpiring in the kitchen after she left them. Phrases like "cheese slipped off her cracker," and "hormonal, right?" and "She's ticked because you asked her to *sing*? Isn't that what she *does*?"

Lucy wouldn't even be there to help Olivia unload her car. Mom of the Year. At the very least, a finalist.

She would have stayed longer in the shower, but the hot water cooled to tepid, as it usually did. Olivia must have filled the washing machine and run the dishwasher at the same time. Maybe Charlie would finally agree to find out why their water heater wasn't keeping up, now that his new business folded before it began. Now that three people depended on it.

Emotions. No linear component at all. She towel dried her hair and slipped into summer pj's, disturbed that she felt both euphoria over her husband's allergy and yet grieved over the loss of the closest thing he'd had to a dream, short-lived as it was.

Their room-darkening curtains and blinds helped minimize the awkwardness of her climbing into bed before dusk. This was not the real her. Not at all. Locking herself in her bedroom like a petulant teen wasn't going to help her find herself. It hadn't helped in middle school.

Her last coherent thought was wondering if either of her housemates realized the basement still sported a gross of unwanted and emotionally toxic worms.

Lucy woke when the bed moved.

Charlie nested like a dog, trying first one side, then the other, then a different pillow, and eventually curling into the same position every night. By the time he finished nesting, every synapse in Lucy's brain was firing. She could go into the family room to read, but Olivia might still be up, eager to talk. Love made Lucy long for that. The ache in her whole being rebelled against the idea. This one night.

She slid her arm from under the covers and patted the shadowed nightstand illuminated only by the face of the alarm. She located her earbuds and the play button for her "Go to sleep, little Lucy" music.

The first strains—slow, simple, honey sweet—quieted and soothed as they always did. Her favorite version of a haunting Bach piece played on a pipe organ in Helsinki, Finland. She'd used it often in the classroom to show the power of a long, slow crescendo—building, building, building in intensity for all five-plus minutes to a glorious high, then ending with a whisper of sound.

She'd used it, too, to illustrate the power of music to express emotion. Few of her students missed connecting with the sense of agony and beauty intertwined.

She didn't need music to make her think. She needed it to bring oblivion back. Sleep. But the notes pulled her along their journey. Each one seamlessly connected to the next. Lucy could visualize the Finnish woman at the pipe organ moving her skilled fingers across the keys, leaving no gaps in sound until the poignant moment in the piece where an intentional rest lifted her hands from the ivory keys of the massive pipe organ and breath itself stopped. Then the music flowed again to its conclusion.

In the video she often showed in class, the organist remained so intent on the flow, on coordinating foot pedals and four rows of keys, so caught up in expressing rather than playing, that an assistant manned the organ stops in sync with her thoughts when the piece called for the sound to change.

Tears dampened her pillow. She hadn't grabbed for music so she could cry herself to sleep. It roared in her ears now with the kind of resonance she could feel in her ribs.

Why did this one have to be the first on her playlist tonight? This one? *Komm, Süsser Tod.* "Come, Sweet Death."

Come, sweet death, come, blessed rest!
Come lead me to peace
for I am weary of the world...

Unlike the unknown lyricist, she wasn't in a hurry to leave the world. Lucy wanted the world to start making sense. She wanted to work, not rest. But that word, *peace*...

She skipped to the next selection on her playlist. Her ears heard a South African children's choir. Worshiping. Much, much safer. Or was it? Their young hearts abandoned themselves so easily to worship.

"This is impossible!" She'd flung the covers aside to sit on the edge of the bed before thinking about the likelihood she'd woken Charlie with the action or her outburst. She pulled the earbuds from her ears and turned toward him to field his inevitable, "Honey, what's wrong?" question.

He snored away like a left-most pipe organ foot pedal. More rumbling than melodic.

She crept out of bed, as if stealth mode were necessary. What time was it? Ten something? Their night owl, Olivia, had to be awake. But as Lucy left the bedroom and closed the door behind her, she saw no lights other than the night-light in the hallway.

The window seat in the kitchen seemed a better hiding place than out there in the open family room with its cathedral ceiling. She'd snatched a comforting but not-too-warm-for-a-summer's-night cotton throw from the end of the bed on her way out and wrapped herself in it. It made climbing into the window seat clumsy. She pressed her back into the side of the window alcove.

The backyard path's solar lights gave the illusion of an inviting haven. But Lucy could see hoards of tiny—and occasionally larger—bugs dancing in the light they emitted. She'd stay where she was. Maybe make a cup of chamomile tea later.

Some people never get a chance to do what they love to do. Some people expend all their energies trying to pull free from the need for government assistance. Some are mired in marriages with miserable spouses. Some live under the thumb of a ruthless ruler or in the shadow of a belching volcano or in flood zones or with sanity-shredding in-laws.

Lucy had worked for almost twenty years in the sweet spot of her passion. She'd raised two exceptional kids to adulthood and shared a bed with a man who adored her, even if he wasn't always sure how to show it.

Then why did she feel as if she'd had an appendectomy without anesthesia? Why did she see fog where others saw clear days? Why did the music that once fed her soul now sting? And why could she not imagine a single scenario where she could reclaim the joy she wore like a signature scarf before?

Like a song that had no end, her mind cycled from sadness to guilt to grief to despair to shame for being stuck in sadness and guilt and grief and despair.

She held her hands in front of her and imagined them raised in readiness in front of her band or choir. Breathe. Downbeat, two, three, four, making a clear point at every beat, as her father had taught her. No mistaking where the beat started or ended. Her fingers poised like a dancer's, even fingertips feeling, engaged with the music, she directed two measures, elbows high, no one watching or responding. Her hands fell into her lap. Useless. Silenced.

Resolved. After Olivia settled in, Lucy would get a job—any job. She'd acquire that counselor, or medication, if that's what it took. And she'd get on her knees. Maybe not in that order.

13

Lucy wouldn't have lasted a week at the public library, even though they were looking to hire someone to shelve books. The forced soundlessness would have pushed her closer to the edge, counterproductive to her goal of holding onto sanity. And how many books would have gotten shelved if curiosity stopped her to take a peek at each one? The arts are the arts.

Olivia needed a job, too. Mom and daughter had compared notes from Internet searches and grapevine buzz around town. "Part-time that pays more than minimum wage," Olivia had said. "Is that too much to ask?" Willowcrest wasn't where she intended to land for long. But commuting much farther than Woodbridge would eat up her income fast.

"Maybe Sam has an idea. Could you work part-time for him? Share his apartment?"

"He has a girlfriend now."

Lucy had looked up from her laptop search. "Why am I the last to hear these things?"

Olivia shrugged.

Lucy didn't want to ask why that might prevent Sam from letting Olivia stay at his apartment.

"It's not what you think, Mom. His life is just a little busy at the moment."

Maybe Lucy should have rearranged the order of her self-help endeavors. Maybe the therapist should have come first.

At breakfast the next morning, after several job suggestions from both Olivia and Charlie—ideas Lucy nixed immediately—in unison they asked Lucy, "Tell us what you're looking for." It would have been cute, their perfect duplication of each other's thoughts, if the question had a legitimate answer.

"I'd have to think about it."

Charlie leaned across the table. "Great. Let's do that." He folded his hands and directed a child-on-Christmas-eager face toward her. Olivia looked just as eager. But then she sat back, wiped a spot of coffee from the corner of her mouth, and changed her facial expression. She laid her hand on her dad's forearm. "Mom thinks best alone. Don't you, Mom?"

God bless you, girl.

"Alone? No," Charlie said. "She's a people person. We'll brainstorm together and come up with way better ideas." His face shifted, too. A boy in trouble. "I mean, two heads are better than one, right? And we can dump all these ideas out on the table and sort through them like we used to do with jigsaw puzzles."

Olivia's grin barely moved her mouth. She stuffed a loose section of her messy ponytail back into the mess. "The thing is, Dad, it's her puzzle."

With his open-palmed hand gesturing back and forth in the air between them, Charlie said, "But we're a team. That's what marriage does to people, Olivia. We're one person."

"Your marriage is the thing that's one person, Dad. That's different. You didn't stop being you. Mom didn't stop being who she is. It's her puzzle. Maybe she likes to work on the edge pieces before she asks other people to help with the middle."

Charlie looked at Lucy. She would have responded if her throat were working.

"And maybe," Olivia added, pushing away from the table to refill her coffee cup, "we're getting in the way of how she best hears from God, so she can even find those edge pieces."

If Olivia hung around for a while, it was possible Lucy wouldn't need an appointment with a counselor.

Charlie thought for a minute. "Then what she should do is—"

"Dad. I know you're trying to help. She knows that, too. Right, Mom?"

Lucy nodded, proud of herself for managing that much without tears.

"The 'shoulds' aren't helping," Olivia said. "The whole town told her she should look for another teaching position in the county or start her own music school or teach piano or voice lessons or just do nothing and enjoy it..."

Charlie raised his hand. "That was me. Except for the money thing."

"The shoulds are like massive perfect storm waves to someone who's drifting. They shove her here and there so she can never get her footing to find out where she's really supposed to be."

Lucy let out the breath she'd been holding. "Psychology. You're in the right field for your gifts, Olivia."

The twinkle in Olivia's deep brown eyes set in that porcelain doll face said something before she spoke. "I think so too. Once I finish my masters, I may keep going."

Charlie patted the table like a final tympani crash. "There you go! See how the ideas just roll when we're talking together?"

Olivia shook her head, still smiling. "Dad, the idea came from me, when I got alone and quiet enough to listen, really listen, to what my heart and instincts—and God—were telling me. What I got here at the table just now was confirmation. And it's huge, having that from the two people I most respect."

"You're a wise young woman." Lucy reached to squeeze her hand.

"Get it, Daddy?"

He stood. "Not really. What I get is that you two don't need me in the picture. You can do very well on your own." Charlie took his cup to the family room. The next sound was the thump of the recliner footrest, followed by the sound of applause. A morning talk show.

Both women shook their heads.

Lucy nudged her daughter's arm, "Go talk to him. You're the psychologist."

Olivia chuckled under her breath at the years-from-now projection. "You go. You're his wife."

"He might need time to think about this."

"I agree."

Lucy took her coffee mug to the sink, rinsed it, and set it to the side for later use. "I think I'll go work on the flowerbed out front. I 'hear' a little better there."

"And I'm going on the computer. I'd like to see how many of my post-grad classes I can do online."

"Honey?"

"Yes?"

"Don't forget to look for school loans for people in your position. Maybe there's something..."

"Good idea."

Lucy removed her gardening gloves and retrieved her phone from her pocket. She opened the "Notes" app and added another thought. "Want to impact lives in a positive way."

She moved to pocket her phone but stopped herself and added, "Look up verses and quotes about singing, song, music." She couldn't afford to discover music—or teaching—had become an addiction rather than a God-given passion. What if her truest role, her truest self, lay in using music to motivate or heal or soothe or...

"If it's all the same to You, Lord, I'd rather not write jingles for products I despise. Other than that, I'm wide open."

Was she?

What if this space between jobs formed not a half rest or whole rest, a measure long, but a tacet passage—an extended stretch of silence? Long enough to lay her instrument down? Or leave the stage? Tacet. Tacit. So similar.

She'd said she was wide open to anything. Really?

So, the job search had turned into a soul search.

Lucy pulled her gloves on and picked up her trowel. Once she got the stubborn weeds out of the way, she could lay fresh mulch. The perennials and the few annuals she planted along the edge would flourish. A bead of perspiration followed the path of her spine down her back. She grabbed a second kneeling pad from her gardening tools basket and created another buffer layer between the ground and her knees. Arching her back, she heard a rhythm of pops and crackles like rice cereal in a milk shower.

As rewarding as a refreshed, weed-free flowerbed would be, she couldn't skip the hard part. Weeding out what didn't belong.

She sat back on her heels and pulled off her gloves to make another note.

The small garden flag Olivia gave her for her birthday the year before waved from beside the airy astilbes. Against a backdrop of a watercolor scene with a bird, a bee, and a butterfly were the words: "Whatever you do, whether in speech or action, do it all in the name of the Lord Jesus and give thanks to God the Father through him," from Colossians 3:17.

Whether vocal or instrumental. Whether in music or...

All.

The dwarf zinnias huddled in front of the flag nearly covered the "and give thanks" portion of the message.

She made a note to purchase a taller flag post.

"Lucy? Phone call for you on the landline." Charlie stuck his head through the opened turquoise door.

"Who is it?" The words came out funny because of the effort of a quick shift from kneeling to standing.

"Bernie."

"*The* Bernie?" She removed her gloves again and took the phone from Charlie's outstretched hand. "Hello?"

"Lucy. Is it true what I've been hearing?"

She sighed. She'd have to tell the story about the RIF again. "What have you heard?"

"That you're looking for work?"

A small portion of the whole story. "Kind of. Yes."

Lucy mentally scrolled through the list she'd started—To bless, encourage, serve, inspire, make their day brighter, to influence children in positive ways, to...

"I have an opening if you're interested." Bernie must have given up small talk for Lent or something. Next year's Lent.

"Waitressing?" Lucy watched Charlie's face scrunch before he turned away from the conversation and headed back into the house.

"Only as a fill-in. I suppose you heard we bought the storefront next door."

She'd steered away from small town gossip. Bernie bought the storefront next door? Maybe her efforts to stay out of the public eye had made her more hermit than she realized.

"We're busting out the wall between and are setting that area with a low stage, a few lights, small tables, and chairs," he said without waiting for a response from her. "Open mic. Recitals. Comedians or semiprofessional musicians coming through here, if we can afford them or they need the gig. Family-oriented. Not the coarse stuff. We don't know what all is possible yet. Or what will fly in this community."

The community that killed music. Or stood by and let someone else kill it. "What would you want from me?" *Whatever you do, do it all in the name of—*

"You know me, Lucy. I sing like a badger with his neck caught in a tangle of barbed wire."

She hadn't completely processed that word picture when he added, "And I have no ear for what's even worth listening to."

"Then why the drive to develop a stage area and bring in talent?"

"Business, Lucy. Well, and community. I think it'll bring in more business. And I think Willowcrest needs something like this. You should hear the grumbling going on over what the school board decision did to the kids... and to you."

She sat on the front step and rested her elbows on her knees. "I don't want to be part of a revenge plot, Bernie."

"No. Nothing like that. Just don't want to see culture completely die off here. With the high schoolers all shipped to Woodbridge because of the consolidation, people have to *leave* here for concerts and plays. Hang on a minute. Judy! Table four is yours. No, I'm sure. Well, check again."

Lucy waited while the difference of opinion ironed itself out.

"So, as I was saying, I can't pay much more than you'd make waitressing for me."

"Bernie, I haven't worked for you in thirty years. I don't even know..."

"Small detail. It'll come back to you. Muscle memory."

That's not what she meant. Wait a minute. Bernie knew about muscle memory?

"I need somebody like you to manage the coffee house part, Lucy."

"Coffee? You hadn't mentioned that."

"We'll have a small coffee bar in there, but we hope people will also want to order sandwiches and fries. Fried cheese curds. Shakes."

Culture. Yep.

"Wish that building had come up for sale sooner. It would have been nice to catch the Fourth of July crowd. I think the best we can hope for is Labor Day. If then. People will start coming back inside in the fall anyway. So, are you up for it?"

"You still haven't told me what I'd be responsible for."

"Everything but the coffee and food."

"I'm not sure what you're asking. To manage the whole project?"

"Line up the live music. Schedule events. Figure out a decent sound system. Invent things like that woodwind quintet you had some of your eighth graders do at the spring concert. That was kind of cool. Or poetry junk. Talent contest. Family stuff. Oh, and Christmas. We could do a lot of things around Christmas. Never too soon to start planning that."

"On a waitress salary?"

"It's all I have for you right now until we see if this takes off and brings in paying customers. You could do some of the planning from home. Just keep track of your hours. And don't overinvest in that department. We'd have to set a limit."

Another restaurant crisis called him away for a moment. Lucy had expected a sense of reignition when the right opportunity crossed her path, the one thing she could get excited about again. Was this it? Her heart stayed mired in neutral, for some reason.

"I'm back," Bernie said. "So, what do you think? I need to cover the afternoon shift starting tomorrow. We've had a sudden opening in that department. But you wouldn't be waitressing often. Not the plan, anyway."

Tomorrow? "Bernie, can I call you back in a couple of hours? I want to talk it over with Charlie." *And see if my heart really wants to get involved in a project like this.*

"He won't have a problem with it, I'm sure. He's the one who suggested you."

Clouds. Scudding across the sky. Blocking out all light.

Charlie begged Bernie to give his poor wife a break? Oh, sure. He meant well. He always meant well. Even when driving her crazy. Well-intentioned crazy driver. *Charlie, come on.*

"Can you do me a small favor and fill in tomorrow afternoon no matter what?" Bernie asked. "Then you can look over what we've got going and decide if this is where you want to plant your feet."

14

"Thanks for fitting me into your schedule, Dr. Hanley." Lucy noted no leather couch in Dr. Hanley's office. The only seating, two French provincial chairs—antique white with plush wine-red velvet upholstery—faced each other near a small fireplace filled with flowers rather than flames.

"I'm glad I had an opening, Mrs. Tuttle."

"Lucy, please."

"Then call me Verna." The lithe woman lowered herself into the chair facing the door. Lucy took the other. "I see," Dr. Hanley said, "that we have the results of your blood work back. Your primary care physician scanned them for me so I'd have them in time for this appointment."

"Great. I think."

"It's good news." A counselor with warmth in her melodic voice. They were off to a good start.

"Meaning...?"

"As a starting point, your calcium is a little low, but menopause isn't your issue. The levels are exactly what we'd expect at this stage."

"So, I can't blame that."

Verna Hanley looked up from the paperwork in her hand. "And neither can anyone else. It's more than a little frustrating to have spouses, children, friends try to pin every quirk or aggravation or outburst on hormones."

Dr. Hanley, for all her youthful appearance, spoke as if well-acquainted with the concept. Beyond the textbooks on her shelves.

"So..." she said, "I'll start the discussion with you the way I do with all my clients. I don't ask the typical, 'Tell me what's bothering you' or 'Tell me what's going on.'"

Good. Because I'm not sure I can figure out how to express that in words.

"Instead, tell me the last time you felt a song in your heart."

"Are you part Native American?"

Dr. Hanley raised her beautiful Nigerian eyebrows and smiled. "No. Why do you ask?"

"Never mind. The last time?" Lucy stared into the depths of the fireplace flowers. "I almost always have a song playing in my mind. I used to have a hard time sleeping because I couldn't find my brain's off switch."

"And now?"

"Sleep isn't any easier. But it's because the music stopped."

"Do you remember the moment when it stopped?"

Let's dive right into this. No dancing around the subject. "I taught music for almost twenty years. Kindergarten through eighth grade."

"How wonderful. Music has always played an important role in my life. I trace that to a third-grade teacher who helped me learn how to use music to express myself when I didn't have the words."

Exactly my point. "My job was eliminated."

"Oh."

"The whole program was cut." Lucy didn't find it any easier to spell out the problem after all these weeks explaining. "Do you have children?"

"Four."

"Do they go to school here?"

Dr. Hanley chewed at her lips as if they'd grown crusty. "No. They're with my husband. In Africa."

Lucy doubted she'd ever hear the story behind the longing in Verna Hanley's eyes. "I wondered if they'd attended Willowcrest. I didn't recognize your last name from the student roster."

"No."

"Well, the music and art programs were cut at the end of this last school year."

"And how do you express yourself now?"

That's not at all the question Lucy expected to hear. "What do you mean?"

"Was music not your own form of expression, too? You say it's been a while since the soundtrack you most loved played in the background of your life. True? Was it that day when the song of your heart stopped?"

"Instantly."

Dr. Hanley took her turn staring into the fireplace arrangement. Unhurried. "The day my husband and children were turned away at the airport in Abuja, their visas confiscated. . . . I mark that as a day the rhythm of my heart changed. They're still in Nigeria."

"I can't imagine."

"Yes, you can. I know you can. You've recently had many, many children who've been told they cannot cross to where you are, who are not allowed to learn with you."

A friend from church had recommended Dr. Hanley, which confirmed Lucy's physician's suggestion. The counseling session wasn't at all what Lucy anticipated. But if all she was able to afford was this one session, it was already worth it.

Someone understood.

"May I ask why your family hasn't been able to join you?" Did the counseled get to ask questions like that of their counselors?

"Entering America isn't the problem. Leaving Nigeria is. We have hope their paperwork will be cleared one day soon."

Lucy tried to imagine being separated from her loved ones by that many miles and a tangle of red tape.

"I have hope for your marriage," Dr. Hanley announced, as if they'd been talking about marriage.

"Well...good."

"You've had your moments of wondering, haven't you?"

Lucy clicked her thumbnails together. "Charlie and I are committed to one another. He's a good man."

"And yet," Dr. Hanley said, "he is sometimes most impossible."

Lucy was paying for this insight?

"And sometimes," the counselor continued, "the impossible one is you."

Now, she was getting her money's worth.

Lucy confessed it all. How she often found endearing things irritating. How she knew she should be grateful for a husband who wanted nothing more than to be in the same room with her 24/7. How she worried the differences they'd weathered throughout their child-raising years—the whole long list of them—were now a serenity-threatening "issue." How she wondered if the only way to keep peace was to give up who she was at her core. How she couldn't imagine *that* was the right answer?

"May I ask you a personal question, Dr. Hanley? Verna?"

"I may not be free to answer."

Lucy glanced at the Bible tucked among the reference books on the bookshelf to the right of the fireplace. Its spine showed signs of wear. "Are you comfortable if we address...faith?"

Dr. Hanley smiled. "I prefer it. I let my clients take the lead on that point. If that's something you'd like to bring into the picture, I'm more than happy to include it in our discussions."

"How does a woman love her husband well without losing herself? Or is she supposed to? And if she is, then what does she do with...with gifts God gave her...with a passion she believes came from Him? And was it horrible of me to tell my husband I didn't want to raise worms? And—"

"Oh, will you look at the time!" Dr. Hanley said.

Lucy leaned back hard against the velvet.

"I'm teasing. Professional joke. We have time left. But it is a knot in a very fine chain, and it will take a good deal more light if we're to untangle it successfully."

"Did you mentally spell light with a capital L?"

The doctor smiled again. Serenity and wisdom. "Did I?"

Lucy hesitated to talk about her feelings of *home*-ophobia—claustrophobia in her own home. Dr. Hanley's house must have felt like an empty gymnasium, devoid of the voices and warmth and faces she loved. Lucy squirmed when Charlie stood a few inches too close or when he hibernated in front of the refrigerator when she tried to make supper. Dr. Hanley's husband stood a continent and an ocean away by comparison.

Maybe that would serve as Lucy's wake up call—realizing how much better off she was than most of the world. No, it just deepened her guilt. She sat up straighter, listened harder, and watched Dr. Hanley's face for signs of hope.

"I have formulated some conclusions and some homework for you, Lucy."

"Okay."

"You will live through this. Your marriage will emerge stronger than ever. You will find a new way to express your passion. And you will not like how long it will take." The empathy in her eyes softened the impact of her final statement.

"I suppose it's natural if I don't yet have a grip on the confidence you have?"

"Expected. Now, for your homework."

"Due next week?"

"No due date. How would a musician express it? *Rubato*. Yes. A relaxation of strict time."

Nothing about Lucy's first appointment with a counselor had followed a pattern she'd expected. Rubato homework fit well. "What would you like me to do?"

"Join the book club."

"A book club?"

"This particular book club. Here's the contact information for the coordinator."

Lucy looked at the name and phone number on the card Dr. Hanley held toward her.

"And I'd recommend you consider taking the job at Bernie's for now. My sense is that it's a passageway of faith to your real destination. Keep your eyes and ears open."

She nodded.

"I'd also like you to practice a technique I call Hemmed Honesty."

"And that is...?"

"You're familiar, I'm sure, with the Bible's teaching on speaking the truth in love?"

"Yes."

Dr. Hanley leaned forward. "I find that many women fall into one of two camps. They speak truth in half-love or speak half-truth in love. What they say may be true, but it is said unkindly or sarcastically. Or they excel in love but aren't completely honest about their feelings and desires. Neither is healthy for a relationship. You can see why."

"I try to be honest with Charlie." Sometimes.

"Hemmed Honesty. No ragged edges. No frayed places in what you say to him. No harshness, but the truth. If he asks where you'd like to go for dinner, tell him."

"He'll choose what he wants anyway."

"Let me describe how it probably happens. Tell me if I'm accurate. Your Charlie may ask where you want to go for dinner, and you will say something ragged, like, 'Anywhere but The Fish Hut. We always go there.' But you don't at all mean *anywhere*. You've spoken half-truth. Not unlovingly but not completely honest either."

"You've been listening in on our conversations."

Dr. Hanley chuckled.

"So, how do I—?" Lucy traced back through dozens of half-truth or half-kind discussions.

"It will take practice. See how he responds when you answer with exactly where you'd like to eat. 'Charlie, thank you for asking. It would bless me if we could eat at the Red Mill. So many good memories there. Would you like me to call and make a reservation?' Did you notice the tone of voice I used?"

"Noted."

"With your permission, Lucy, I'll be praying for you. You've borne a grave disappointment. And waiting takes courage in larger measure than most people understand. Courage to you as you wait."

She said it as one might toast the eve of a battle.

15

How had she misplaced the note from Ellie's mom so long? She remembered having received it, but when? The postmark said it had been mailed more than a week ago. It stood today leaning against the coffee maker.

"Olivia, did you put this here?"

Her daughter tugged at two halves of the ponytail at her crown in an effort to make her morning self more presentable, Lucy assumed. Oh, to be so young that's all it would take.

"No. What is it?"

"A note from a student's mom."

"Must have been Dad."

"I wonder where he found it." Lucy paused. "I wonder where he is." He, the one who had gone behind her back to talk Bernie into hiring her. Somewhere in the curl of anger swirled a faint flicker of gratitude that Charlie cared. Excessively.

Olivia opened her laptop on the kitchen table. For her, breakfast consisted of coffee and the ever-present job search. Lucy resisted saying, "Won't you at least have an English muffin or a piece of toast? A smoothie?" as she would have if Olivia were fifteen.

Lucy held the envelope in her hand, mentally apologizing for neglecting it so long. She slid a finger under the flap, suddenly overcome with a deeper remorse. What if Ellie had been hospitalized again and Lucy had ignored it?

"I'm taking my coffee out to the deck, Olivia."

"Go ahead. I'm trolling for a phenomenal career opportunity within driving distance. Yeah. I'll be here a while."

The deck welcomed her as if it had a gift of hospitality. Aromatherapy. Cloud-free day so far. A bright spot caught her attention near the bird feeder. Joined by other bright spots. Goldfinch convention in progress. Other birds waiting for their turn at the seed buffet peppered the maple tree nearby and chirped their impatience.

"Easy to hum," she told the aviary songsters, "but not easy to dance to. I give it an eight."

A handful of finches bounced on the roof of the bird feeder as if to refute her statement. Such comics.

She retrieved the towel she'd slung over her shoulder for the purpose and wiped the last hints of dew from the patio chair and table.

Even on the warmer days during the school year's early fall and late spring, she saw so little of this sweet deck, the small but mighty vegetable garden, the stretch of lawn Charlie kept trim with an even greater diligence now that he was retired.

The deck and landscaping were situated so this part of the yard seemed tucked off by itself, secluded despite its closeness to other neighborhood lawns. A haven. Okay, so maybe there were a few benefits to having more time off.

Lucy pulled the note card from the envelope.

Dear Mrs. Tuttle,

She scanned ahead to catch any mention of a decline in Ellie's health. No. It looked like a thank-you letter.

> *Ellie and I can't stop talking about how empty this summer feels without the private lessons you used to teach. Other young people probably have many activity options to choose from during the summer*

break. Ellie's one passion was the private music lessons with you.

We've both noticed a regression in her breathing. I don't say that to make you feel guilty in any way. But to let you know that you have had a profound effect on my daughter's life and her enjoyment of it. She still plays, on her own, and sometimes will play along with a YouTube video for the discipline of blending and timing, breathing...

Personally, I wanted to express our gratitude for providing not only a safe haven for her, but for igniting a passion that has made her brave in so many more areas. That hasn't been easy, because of her CF. You've served as both influence and inspiration. We miss you already. Please let us know when you find your next position. Ellie and I joke that we'll move the whole family there, even if it's out of state.

God bless you as you seek Him for the glorious What's Next?

A go-cart? A drone? What was that noise? Charlie's weed trimmer. At this hour of the morning? If he didn't care about disrupting Lucy's peace, he should have at least cared that others in the neighborhood might still be in bed.

Blaze orange, heavy-duty earplugs sticking out of his ears, Charlie waved from the base of the maple tree. Birds scattered. Lucy could almost feel their pea-sized hearts beating out of their feathered chests. Charlie revved the weed trimmer's engine, apparently a requirement, but it reminded Lucy of a motorcyclist revving purely to prove his engine's—and his—firepower.

So much for serenity.

She grabbed the note and her coffee cup and retreated to the kitchen. Olivia's laptop was now in her lap and her feet propped on the kitchen table. Lucy refilled her coffee and moved to the family room. Olivia giggled. Lucy hadn't been in her chair a whole minute when Olivia said, "Mom, you have to see this. Stay there. I'll bring it to you."

The "have to see" was a YouTube video of a cat trying to jump from a radiator to the top of a dresser. Failing every time. How many million views? If Lucy hadn't been going through whatever crisis this was—Dr. Hanley hadn't defined it for her yet—would she have found it funny?

"There's another one," Olivia said, searching for another link.

"I'll have to skip it this time," Lucy said. "I have book club this morning."

"This early?"

"Errands to run first. I'd better fix my face."

And by "fix my face," she meant relax her jaw, unfurrow her brow, and find where she'd misplaced her joy. She had no glorious "What's Next?"

She could easily waste time at the greenhouse looking for a flat of annuals to replace what the chipmunks had chewed off in the night. Or rabbits. Charlie hadn't objected to the idea of Lucy attending a weekly book club, although she'd seen in his eyes a brief flash of a young boy's "Can I come, too?"

Or was she reading him wrong?

After a half hour in the greenhouse, she realized any plants she purchased now would have plenty of time to bake to a lovely crisp brown in the car during her first meeting of the Hat Club book club. It also occurred to her that her earlier hasty exit from the house had made her forget to grab a hat. Her office shelf held a couple of good options—costume hats she'd worn for various music time period or style units at school. Rather than return to the house and risk being late for her first time as a book club newcomer, she purchased a lime green sun visor near the check out counter at the greenhouse. It would have to do.

Per Dr. Hanley's instructions, Lucy parked her car in the city lot behind a row of historic shops downtown. A small brick courtyard formed the back entrance to a used bookstore that smelled of coffee and dusty tomes. Just inside the back entrance, a hallway led toward the body of the store. An archway to the left mid-hallway opened into a room Lucy hadn't noticed before. Empty except for a collection of four mismatched tables and even more mismatched chairs gathered around them, the room looked like an ex-storeroom converted into a think tank. At one table sat three women—younger than but reminiscent of TVLand's *Golden Girls*, none of whom Lucy knew. None wore hats.

"Are you looking for us?" a Bea Arthur look-alike asked.

"I must have the wrong room." Lucy removed her lime green visor. "I'm looking for the"—she indicated the item in her hand—"Hat Club book club?"

Bea smiled at Betty White. "That's us. You must be Lucy. Come join us." She used her foot to scoot a high-backed chair away from the table they occupied.

"The Hat Club?"

"Common misconception. Not that we mind. Makes it easier to announce where we're headed," the fortyish version of Betty White said. "We're the HHATT club, spelled with two h's and two t's."

Lucy took the chair offered. While she mentally wrestled with the reason to spell *hat* with two h's, the third woman spoke up. "HHATT. He's Home All The Time." She emphasized the last three words, but finished with a bright as July smile.

Dr. Hanley, what have you gotten me into?

"We know you're Lucy. Let's introduce ourselves. I'm Marta," Bea Arthur, the ringleader, said. "This is Carole," indicating Betty White. "And the *quiet* one"—she nodded toward the third musketeer—"is Angeline."

"Pleased to meet you. I apologize for not having read the book yet." At the curiosity of faces, Lucy added, "The book we're discussing?"

"We're not that kind of book club, Lucy." Marta elbowed Carole who elbowed Angeline who elbowed Lucy.

"I don't understand."

"Dr. Hanley loves this moment, when a new member discovers our purpose here. We're not reading books, Lucy. We're writing one."

It wasn't her green visor alone that upped her level of discomfort.

Angeline stood and closed the pocket door to the think tank. "We're writing the book"—she said—"on how to survive when your husband's home all...the...time."

Lucy felt compelled to explain, "I...I love my husband. He's a good man."

"Excellent," Marta said. "It's a prerequisite for the HHATT Club. We all love our husbands. We love being with them. It's the 24/7 part that needs a little help. And the part about how an empty nest can exacerbate that dreaded oppositeness that worms its way into what we thought was going to be bliss."

"*Worms* its way? Did Dr. Hanley talk to you about that?"

"Dr. Hanley tells us nothing about new members," Carole said, drawing a hot pink patent leather journal from her tote bag. She waved it like a grand prize in a game show then handed the journal to Lucy. "All we know is what each of us decides to tell." She looked at Marta. "I don't even know your last name, do I?"

"In time, young Grasshopper."

Lucy folded her hands between her knees. This could get interesting.

"Take a look at your book, sweetie," Marta said.

The magnetic closure opened with a small tug. "It's blank."

Each of the other three pulled out similar journals. "Not for long," Carole said and opened hers. She fanned page after page of handwritten entries. "Did you bring a pen or pencil?"

Lucy dug in her purse for her favorite gel pen, the one that wrote with the color and flourish of fountain pen ink. She uncapped it and opened to the first blank page. *Please don't ask me to write down my feeeeelings.*

"So, God," Angeline said, without preamble, "it's us again. We're just as needy as we were last week. And"—she opened her eyes and paused dramatically—"begin. Amen."

"Amen."

"Amen."

"Amen?" Lucy added.

Marta must have heard the question mark. "Angeline's our chaplain," she told Lucy.

Uh huh. Explains... everything.

"I have something," Carole said, bouncing like a first grader with the answer. All eyes turned toward her. "I saw it on a friend's Facebook status."

Oh, dear. Dr. Hanley, this is not—

"She and her husband are celebrating their fortieth anniversary. And I happen to know they haven't had the best marriage until recently."

Marta licked the tip of her pencil, ready. Lucy hadn't seen anyone moisten the tip of a pencil since school scenes in *Little House on the Prairie* reruns.

Carole slipped on cobalt blue reading glasses. "She said that we should all stop, take a minute to evaluate our own choices, and chart a new course of kindness if we need to."

"Say that again, please," Angeline said, "a little slower."

"Largo." Lucy pressed her lips together. She still thought in the language of music. *Largo—slowly.*

"She's more of an *allegro* person," Marta said.

A musician? "Marta, do you play piano? I know that's way off topic. I just wondered."

Carole and Angeline exchanged knowing glances. Carole spoke. "You could say that. She's only the first chair trombone for the Reedman Symphony Orchestra."

Lucy couldn't help herself. She zeroed in on Marta's lips. Trombone lips? Definitely. Marta formed a perfect brass embouchure and buzzed her lips for emphasis.

"Fascinating."

"Are you a fan of woodwinds, strings, double reeds, percussion, or brass?" Marta asked.

Lucy laughed aloud, then stifled the sound. "All of the above."

Angeline leaned forward. "Good answer."

"Until recently"—how far should she go?—"I taught music at the Willowcrest School."

All three of her book club companions leaned back, a perfectly synchronized movement. "Oh," Carole said.

"Oh, oh." Angeline.

Marta sighed. "You're the one."

"The one?"

Carole's sigh joined Marta's. "The riffed."

Where had Dr. Hanley found these people? "You understand the term?"

"Riffed?" Marta raised her hand. "Four years ago."

"Riffed." Angeline. "Twice."

"Riffed." Carole's bounce had grown decidedly *legato*. Smooth. "Best thing that ever happened to me."

Her comment drew disagreeing but not unkind stares from the other two. "Carole," Marta said, "does not share our opinion about the . . . life changes . . . brought about by the RIF process."

Angeline leaned forward again, as if her every word needed to be whispered. "She left a job that made her miserable and in her next teaching position met the love of her life."

"Well," Lucy said, "I already have—" She did. She already had the love of her life. He was home, weed whacking.

"But . . ." Marta interjected, "he's—say it with me—*home all the time.*" The choir of voices ended with a reprise of laughter.

That Dr. Hanley. Smart woman.

16

Olivia met her just inside the door. "Mom, Mom! Do you have any shoes that go with this?" She was dressed in a black pencil skirt and white blouse with black polka dots. "Interview in an hour in Woodbridge, but I can't wear flip-flops."

"Cherry red flats?"

"Perfect. Bedroom closet?"

"I'll find them. Sometimes I have to store extra shoes under the bed."

The two women headed for the bedroom. "How was the book club?" Olivia asked.

Lucy's shoe pursuit paused a hitch. "Great. Really. Good people." Oh, so good.

"Wonderful."

Lucy scanned the shoes on the floor of the closet, then abandoned that search and dug under the bed for the slide-out drawer. "Ah. Here. Hope they fit."

Olivia slid her bare foot into the first shoe. "I'll put tissues in the toes. They'll be perfect. Thanks." She snatched four tissues from the box on the nearest bedside table.

"Where's your interview?"

She mumbled, bent in half to adjust the shoe sizes.

Lucy started. "Did you say, 'The bar'?"

Olivia shuffled around the bedroom. One way to break in a pair of shoes. "The Barre. As in, ballet barre."

"You're interviewing as a dancer?"

"Mom, you're a riot. The Woodbridge Center for the Performing Arts. It's only a receptionist position, but..."

Lucy was two hours away from slipping into her sky blue "Bernie's" polo shirt to serve the sparse afternoon crowd their Reubens and taco salads. If she'd known about the opening at the Center for the Performing Arts.... She stood in Olivia's path, took her by the shoulders, and said, "You will make an exceptional Director of First Impressions." She kissed her daughter on her forehead.

"I don't have the job yet."

"But you do have the shoes for it. Now, go. And drive carefully."

Olivia tucked her blouse into the skirt waistband. "How old will I have to be before you stop telling me that, Mom? To drive carefully?"

"A hundred?"

"Okay then. Praying for me?"

"Of course. Hey, where's your dad?"

Olivia called over her shoulder, "Don't know. See you later."

How could a man who was home...all...the...time disappear all...the...time? Lucy checked the kitchen, family room, and deck. Back in the kitchen, it occurred to her that the basement was the only unsearched square footage. She opened the door and hollered down the stairwell, "Charlie? Are you down there?"

No response.

But the light was on.

Why didn't he answer?

As she started down the carpeted stairs, she held tight to the handrail and to the logic that told her nothing was wrong. Logical explanations abounded. Eighty-five percent of assumed trouble turned out to be a false alarm. Or was that ninety-five percent?

She couldn't see into the basement until she reached the landing at the bottom of the stairs. They'd talked about opening up the wall but had decided it would be safer this way when grandchildren came along. Someday.

They'd talked about framing in another bedroom in the basement, too. Maybe make the sad little toilet and shower into a legitimate third bathroom. Right now, the basement served as open storage. Drywall and concrete floor. A small area for an additional workbench to supplement the space in the garage.

She turned the corner and scanned the unrealized potential. "Charlie?"

He popped jack-in-the-box style from behind a pile of storage bins. "Yeah?"

"Charlie! You scared me."

He disappeared behind the bins again. "Huh?"

"Didn't you hear me calling you?"

He popped up again. "Huh?"

"Charlie. Take the ear plugs out of your ears." She tugged at imaginary ear plugs in her own ears.

"Oh! Forgot they were there. Olivia was watching a goofy movie marathon thing at top volume."

"What are you doing down here?"

"Looking for something."

"What?"

"My wife. She's never here."

"Kidding, Lucy. I was kidding."

He followed her so closely she could feel his exhales against her neck as she pounded up the stairs.

"Lucy. Kidding."

Book club. Book club. Book club. *Choose a ... No. Chart a new course of kindness, if necessary.* She turned to face him. "Charlie, did it bother you that I went to my book club meeting this morning?"

"No," he said. "It bothered me that you didn't say good-bye."

That was his problem? "You were running the weed trimmer."

"I would have stopped to kiss you good-bye."

How could she argue with that?

"What if—?" He stopped mid-thought.

"What?"

Charlie crossed to the sink and washed his hands. "What if you hadn't come home?"

Had he grown paranoid all of a sudden? Is that what retirement did to a person? Oh, no. Was she paranoid about becoming paranoid?

"Charlie, I'll always come home." She hugged him from behind while he attempted to dry his hands on a paper towel. She leaned her head against his back. His shirt smelled like a mix of perspiration, fresh air, dryer sheets, and grass clippings.

He shuffled around until her head rested on his chest and his arms enveloped her. "I hope so."

She tilted her head back to look him in the eye. "You know so. Don't you?" Whose fault was it if he wasn't sure? And why did relationship issues always have to be someone's fault? Whose idea was that?

Humans. Messy creatures. *Clumsy mortals.*

"What were you really looking for in the basement?"

"A reason to live."

"Charlie Tuttle!"

"I was looking for my old fly tying kit. Something to do."

A fissure formed in the lining around Lucy's heart. She'd been sidelined against her will, but clamoring to get put back in the game, longing to discover where she fit, where her love of teaching and her passion for music would take her next, fighting fear every day that those years were over.

Charlie just wanted something to do. Some little thing to putter with.

How could they be more different? Did marriages that survived this season of life ignore those incongruences? Or were the survivors those who formed perfectly matched pairs? In tune in a circadian way with the rhythms of work and rest, activity and inactivity?

"You object?" Charlie said.

"What? No."

"You weren't saying anything. I thought you didn't like the idea of my tying flies." He filled a glass with ice from the refrigerator door.

"Is that how you envision spending the rest of your life, Charlie? With your hobbies?"

He added water to the ice and took a long drink. "Is that a bad word? Hobbies? You said it like a hobby is a big bowl of overcooked lima beans."

Had she?

"And for the record," he added, "I envisioned spending the rest of my life with you."

"But not *just* with me. Right? I mean, you didn't dream about our living out the rest of our days sitting in matching rockers, did you?" *Chart a new path of—*

"Recliners." He set his glass on the counter with enough oomph to rattle the ice cubes and left the room.

And, on that note, Lucy realized it was past time to get ready for work.

Positive side of working outside of her chosen field? Her former students were often her customers...with their parents. After an initial blush of embarrassment, Lucy figured out a routine.

"Hey! Mrs. Tuttle! What are you doing here?"

"Having an adventure. And getting a chance to see you again, Brandon. How's your summer going so far?" Lucy looked from child to parent to see if their answers would coordinate.

Lucy had steeled herself against the seeming community-wide remorse about the funding cuts for the music and art programs.

Pride tugged at her when she realized it felt good to be missed. A little too good. She soon became expert at diverting conversation.

And then she'd produce the inevitable, "Would you like fries with that?"

How many parents asked her about teaching private lessons in the first weeks after her employment at Willowcrest ended? It should have been completely natural for her to accept every inquiry with gratitude. Something didn't sit right. The requests made her restless in a way she couldn't explain. Agitated when she should have been grateful. She did not relish trying to describe the feeling to Dr. Hanley.

How many community members asked if she were still playing for, singing for, or coming to their late summer and fall weddings, as if losing her position meant she'd lost her talent. She hadn't lost her talent. But her interest registered at a decibel level below that which could be heard by the human ear.

She'd committed to the weddings. She'd be there. The wedding she'd accompanied mid-June was followed by two funerals for which she'd sung the next week. She'd lived through them all. Somehow.

For two weeks now, Lucy had tried to pin Bernie down for more details about what he wanted in the coffee house area. A blue tarp hung over the interior doorway that had been created between the two buildings, but that's as far as any construction had progressed.

"Little hitch in the building permit," he'd said. "Keep dreaming up ideas. We'll get over this hurdle soon."

But that was the thing. Ideas were pale and scarce, like sunbathers at the North Pole.

Her lack of imagination, her dearth of creativity scared her. It sounded like another symptom she should probably bring up to Dr. Hanley. She should have been filling notebooks with plans, creating computer file folders, searching the Internet for potential comedic or musical talent. Even if it took several months to finish the construction part, she needed to have talent booked a year or more in advance. The nonlocal talent. Maybe even coordinate with the Performing Arts Center in Woodbridge to share talent needing another gig in the area before moving to their next.

But every idea soured before she developed it beyond a fleeting thought.

Unfinished symphonies. A bunch of them.

It's not that she was idle at home. The vegetable garden had started producing. She'd picked strawberries at the patch outside of town and made jam. She'd tried tackling the office—lots to toss or give away even if she did find her song again. But something had called her away from the task every time. She couldn't remember what.

Lucy hadn't heard from Ania for three weeks or more when she showed up at Bernie's on a day Lucy had planted herself on the other side of the blue tarp, looking for inspiration.

"Charlie said you'd be here," Ania said when Lucy greeted her.

"Good to see you. Let's pull a couple of chairs from the restaurant side. Better yet, let's grab a booth."

"Bad time? I can come back."

Lucy conjured her best imitation of a "Welcome to Bernie's" smile. "Let me get you something to drink, too. What'll you have?"

Ania flipped her hand as if it didn't matter.

"Iced tea?"

"Sounds good."

A minute later, Lucy sank into the booth bench opposite Ania. "You look...tormented."

"Nice to see you again, too."

"Something wrong?" Lucy slid a napkin Ania's way when the iced tea left a puddle on the tabletop.

"I signed a contract today. Teaching art at a charter school in Madison." Ania paused. "It's a...a dream job."

Lucy covered her mouth with her hand, holding back the sob that fought for room to breathe.

"You can say, 'I told you so.' I deserve it."

When she managed to swallow the sob, Lucy said, "I knew you didn't have it in you to walk away from teaching. Not with your obvious talent and gift for drawing it out of others. Not after one disappointment."

Ania raised her eyebrows. "So, it's a disappointment now? When did getting riffed change from devastation status?"

The tightness returned. Ania found a position. Her dream job. "Lucy?"

She beat back the stability-threatening wave of jealousy. "I'm so happy for you. Really." She was. For Ania.

The young woman pulled a package from her burlap tote bag. "This is for you. And don't even think about refusing it."

Lucy hadn't torn more than a corner off the handmade wrapping paper before she knew. "Oh, Ania." The repurposed cupboard door. "Some days there won't be a song in your heart," Lucy read out loud. "Sing anyway."

"I was going to wait for Christmas, but..."

"You knew I'd need it today."

Ania rotated her glass of tea. "I knew I couldn't share my news without it. You're the one who restored my hope."

I'm no hope-whisperer. Lucy ran her fingers along the edge of the gift. *If I were, I'd be talking to myself.*

And now I'm talking to myself.

17

Did you feel your friend held back on her excitement for your sake?" Dr. Hanley set a dry pen aside and picked up another.

"Definitely. As if she felt sorry for me," Lucy said, worrying the tissue in her hand. "And of course, that ramped up my guilt that day. And, okay, maybe it's still there. I should have been cheering louder for her. I should have been the one checking in on her every week to make sure she was okay. She considers me her mentor. Her hope-whisperer. Isn't that laughable?"

"And the guilt came from...?"

Lucy slid her feet in and out of her sandals. The action didn't help her think. "I need to be stronger. I should be weathering this better. I have twenty years of age and experience on that girl."

"Plus that faith thing you have going on."

Was she serious? Or practicing her sarcasm skills? "Dr. Hanley, I can't even find the puzzle box, much less start putting my life back together." *It's her puzzle, Daddy.* Olivia, one of two brilliant children to whom she'd given birth despite not knowing how to guarantee brilliant children.

"Interesting." Dr. Hanley offered Lucy a glass of ice water, then settled back into her chair.

"Thank you for this." Lucy stuck the tissue in her purse to put it out of its misery. "It's not that I don't think you're perfectly capable with puzzles."

"But it's your puzzle."

Lucy didn't remember having shared Olivia's assessment with her counselor. "True."

"And yet...?"

Lucy searched Dr. Hanley's face for a clue. The woman was good at hiding them. "Wouldn't it be interesting," Dr. Hanley said, leaning her head against the back of her chair, "if we could chart a timeline of our range of emotions when a crisis hits. That initial explosion of pain followed by a long stretch of aching, then acceptance..."

"Then another stab of pain."

"Yes. It's not a smooth incline. For any of us. When you spoke with her, you were sitting in a room devoid of inspiration, correct?"

Lucy checked the clock on the wall, though it was early in their session. "Yes."

"Do you think that may have affected your reaction?"

"Emptiness? Barrenness? A room without a cause? Realizing my creative mind can't remember how to create anymore?" Lucy chewed on the inside of her cheek. "What if her new position brought out plain, old, ugly, despicable jealousy in me? What if that's all it really is? I'm not used to that. And I don't like it. Tastes like metal. Dipped in chalk."

Dr. Hanley chuckled. "You don't think that description is dripping with creativity?"

"Wouldn't make a good song lyric."

"You write songs, too?"

"Wrote. Past tense." Lucy chewed on her other cheek while she waited for reprimand.

Dr. Hanley drew a shape on her notepad. From the path her pen took, it looked like a square. Box. A cell. "So, it's over then. You no longer believe music has value. Your love for your students has dissipated. Joy of living is gone."

"That one. Yes."

"Permanently."

"No." Did she really mean that? Did she believe this cloud she was living under would lift someday? And if she believed that, why did she act as if it were a permanent fixture? "I see what you did there. Reverse psychology. My daughter would be impressed. Thank you for that perspective." Lucy's knotted stomach untwisted itself.

"Perspective is my specialty." Dr. Hanley smiled. "I'm glad you keep making appointments, Lucy. Your crisis, crises rather, are still in their early stages. I encourage you not to judge your final outcome by the first walk-throughs of the storm damage."

"Someday, Olivia will serve the role you're serving for me now. I didn't think I could be any prouder of her. But knowing what she's called to do, and the impact it can have..."

"I believe your father, were he still alive, would say the same of you regarding music education."

"I failed him so miserably." Lucy's words barely made it past her mouth.

"All you accomplished in the nineteen years you taught at the Willowcrest School vaporized when you received the letter, correct? It and the students you reached meant nothing. What really would have counted is what you would have done with the *next* year's students, and the next. Is that what you're saying? You're grinding your teeth, Lucy. I can tell from the tension in your jaw. My husband, the orthodontist, would advise against that."

Lucy crossed her legs and dangled her sandal from her big toe. "Is your air conditioner working?"

"Perfectly." Dr. Hanley tapped her pencil on the cover of her leather notebook.

"Do you want to hear something curious? Well, not curious. Surprising?"

"Always."

Lucy stretched her legs in front of her. "Charlie knows I'm coming here, of course. We'd talked about it initially. He was a little too eager for me to start, I thought. When I told him how much each session costs, he suggested we take money out of the savings account. Or get a loan." Lucy laughed. "Savings! At first I thought, 'Am I that sick?' or 'Are you that sick of my moping?'"

"Your attitude changed?"

"He wants me to be well. He knows I need someone like you to talk to."

"Good man."

"But I think he still assumes he should be that person."

"Ah."

Lucy heard music coming from the offices next door. She pointed with her thumb.

Dr. Hanley nodded. "Yes, a new business moved in this week. Drum shop."

"And the soundproofing between the walls is..."

"...about to be improved." Ever gracious Dr. Hanley. "Let's get back to Charlie for a moment. You believe he assumes he should be enough."

"I think so."

"He is your life partner."

"Yes."

"He is not your God."

Lucy startled at the obvious simplicity of the statement. "No. But where do I draw the line in the respect department? I know my responsibility is to respect him in our marriage."

"Responsibility or high honor? And just so you know, there is no line. There is only grace."

Lucy sat back. Stung. "I feel as if I'm disrespecting him if I need time alone. It's the way I'm wired. But if that hurts him..."

"Lucy, you are not his God either. Our husbands, too, are responsible to gain their completeness from Christ, speaking a little more strongly than I usually do with my clients. No matter what we do and no matter how much love and respect we show our husbands, none of that will impact them to the degree God can."

Dr. Hanley's gaze drifted to the mantel. The pictures of her husband and children.

"That helps. It does."

"Our job is to make sure we're not getting in the way of that happening. And that," she said, "will take some time to unpack."

"Hence," Lucy said, "the savings account."

"Package came for you," Charlie said when Lucy deposited six plastic bags of groceries on the counter. "Welcome home."

"Thanks."

"Good session?"

"Dr. Hanley's very wise. And sometimes brutal."

"Great. I'm heading over to Martin's."

Great? He wasn't going to press further? Did he have no curiosity cortex? Maybe she should be grateful for that.

"Eve's with their daughter today so he and I are going to tackle some fix-it projects he hasn't had time for."

Lucy pulled a carton of eggs from the first bag. "You'll be a blessing. You always are."

"And we might be able to fit in an hour or two of fishing."

Lucy dug into another bag and smiled. "It would be good for both of you."

He grabbed his hat on his way out the door. "Don't forget about the package."

She put away three kinds of parmesan cheese. The kind she liked—shaved. The kind Olivia liked—shredded. And the kind Charlie liked—pulverized. If purchasing groceries became a problem before their reserves ran out or Lucy found a position with legitimate earning potential, they might have to narrow it down to one kind. She wondered if a marriage had ever broken up over parmesan cheese.

Groceries led to kitchen clean up, emptying the dishwasher, taking out frozen chicken to thaw for supper, chicken they wouldn't need if Charlie brought home fish.

The small package waited for her on the hall entry table. She hadn't ordered anything. And the box had no postmark. She tore off the packing tape far enough to get her fingers between the top flaps of the box. Charlie, Charlie, Charlie. New garden gloves. Worm print. With lettering on the back of each one. "You've wormed your way," read one. "Into my heart," read the other.

The romance never stops with you, does it, Charlie?

She nestled the gloves between the layers of tissue paper. *You've wormed your way into my heart, too.*

Lucy had hung the small cupboard door from Ania in her kitchen, in the window seat alcove. Her meditation spot. She considered repainting the frame to match the décor. But the sentiment had a permanent spot in her home, even if it did struggle to nest in her heart.

Charlie would be gone for several hours. Olivia had a second interview at the Performing Arts Center. Lucy grabbed her pink journal from the desk drawer in her office and retreated to the window seat.

She started a new page, taking a cue from something Dr. Hanley said just as their session ended. "Lies I've Too Long Believed."

After she'd written the heading, nothing else came. She had a feeling if she waited long enough, something would show up.

Maybe if she made some sun tea.

After she'd set the tea jar on the deck, she noticed how tall the radish leaves had grown since the last rain. Maybe she should pick radishes for their supper. And thin the carrots. And check the beans for blossoms.

Instead, she retreated to the window seat and her journal. She'd thought of the first lie.

"What I did all those years didn't matter."

After the first one was committed to paper, more showed up.

"Job loss equals job failure." She knew that wasn't true. Knowing, believing, and acting are different things.

She made another dot for a bullet point: "Dad would be disappointed in me. Heartbroken." Of all people, her father understood the never-ending budget battles between music departments and school boards. It had been a frequent point of dinner table discussions. He'd been blessed with an advocate on the school board for years. What might disappoint him is if she never recovered.

"Needing professional help is a sign of weakness." She leaned her head back and laughed at that one. How many sixth graders would pick up a clarinet and stiff-arm her with 'I can do this myself. I don't need professional help'? And how many of those would still be in band in seventh grade?

"Suffering together isn't intimacy."

What had she written? Where did that come from? She stared at the words on the page in front of her, considering what they might mean. She wrote them, but didn't recall thinking the thought. Or had she? Lucy had tried to keep the agony part to herself, thinking that would help preserve her marriage relationship. What if the opposite were true? What if talking to Charlie about how she really felt would have a positive effect on their relationship? What if suffering *together* were glue?

"The second half of life is just survival." She almost scratched out the words, but left them.

The next few came as lie suggestions from Dr. Hanley: "Your spouse is the enemy, not a gift." The woman had said she was merely offering examples, but each one stuck.

"Unequal passion is the same as unequally yoked." Another lie that felt a lot like truth at first glance.

"A dream that changes shape is a busted dream, not a grown-up dream." Lucy scratched out *grown-up*, but couldn't think of a fitting substitute, so she rewrote the phrase.

"Differences make us incompatible."

She had radishes to pick. The tea might be ready. Or she could stare that lie in the face and see it for its true colors.

18

Fish for supper. Radishes and early peas on the salad. Fresh brewed sun tea.

"I love my new gloves." Lucy wore them to the supper table to prove her point.

"Stylish."

"Where did you find gloves with an earthworm design?"

He opened his mouth to answer.

"Wait. Maybe I don't want to know."

"Hypoallergenic." Charlie poured Lucy's homemade balsamic vinaigrette over his salad.

Oh, no. "You found a breed of worms that are hypoallergenic?"

"Oh, to be so lucky. No. The gloves are."

Lucy set the gloves on the other side of the table to keep them from getting dirty, as if they wouldn't the first time she used them in her flowerbed. "Does this mean you've forgiven me?"

"For what?"

"For not sharing your interest in worm farming?"

"Nothing to forgive," he said. "Besides, I'm pretty sure I could have won you over, if I hadn't been allergic."

Pretty sure you couldn't.

She should have anticipated it, but didn't. After supper, Charlie headed for his recliner. Baseball game. Loud, loud baseball game. Like every spring and summer.

It's a lie, she reminded herself, *that differences make us incompatible. A lie.*

But there was no getting around it. She'd either have to find noise-canceling headphones or leave the house. No corner was immune to the constant crowd noise and the occasional eruptions by Charlie regarding what he deemed either a great play or a bad call. Volume was the same for either.

"I'm going for a walk, Charlie." She tied on her athletic shoes.

"Why don't you sit here with me for a while?"

Because I'll go insane. Other than that... "Is it okay that I don't care for baseball like you do?"

"I'll turn it off, then, and go for a walk with you."

"Then I'll feel guilty for pulling you away from the game."

"I'll put them on the DVR. Watch them later."

"Them?"

"Double-header."

Record them. Which postpones the noise, but doesn't eliminate it. Noise and music shared practically nothing in common.

The landline rang, interrupting the riveting discussion. Bernie. Someone called in sick, though Bernie doubted the story he'd been handed, he said. Was Lucy free to fill in? Yes. The answer was yes. But as a courtesy, she should check with Charlie.

She held her hand over the microphone end of the phone. "Charlie, Bernie needs a fill-in. Any objection?"

"Will you look at that? Double play! Way to go!" He clapped his hands as if personally responsible for all the crowd noise in the stadium.

"Charlie?"

"What? Oh. Fine. That's fine."

Differences do not make us incompatible.

She should have chosen the ball game.
Ball games. Plural.

The malt machine broke down, mid-malt. A catsup bottle exploded, catching the other waitress in the face and Lucy on her new, super-supportive shoes. The sweet tea and unsweet pitchers got mixed, causing no end of—as her Louisiana roommate from college would have said—*consternation*. Sweet tea hadn't even been offered in Wisconsin until a few years ago. Waitressing hadn't been this hard back in Lucy's early twenties.

And finally, a family walked in and asked if their oldest son's meal could be pureed, since he'd just had extensive dental work.

He ordered liver and onions.

As she worked on her responsibilities at the close of the shift, her eyes drifted often to the blue tarp hanging between two futures. On this side—her current life—nothing was as she would have hoped. She would have thought the other side—even though unknown—would have called to her. Shouldn't she be lying awake at night with ideas tumbling over each other for attention? Shouldn't she itch to pull back that drape and dream about the possibilities? Shouldn't creativity have kicked in already?

She no longer pestered Bernie for information or updates. She'd slipped into a work existence—the few days she worked—that marked success by turning seared cow organs into a puree fit for a dental patient.

Was it arrogance that made Lucy so picky? Why hadn't she accepted the offer to serve as music director at their church? It matched much of her skill set.

It didn't match her passion. It's not that she didn't appreciate worship. Some of her most meaningful moments spiritually had been connected to music. What was she waiting for? Writing on the wall? An apology? A divine apology?

No. She wasn't that shallow.

She'd been given a gift for nineteen years—to do exactly what she wanted to do. Every day. It was hard, taxing, heartbreaking sometimes. But it was exactly what she'd hoped to do with her life.

With one letter, all that changed. "I'm sorry. Your services are no longer needed." Anywhere.

Even before she was hired by Willowcrest School, she'd done exactly what she'd longed to do—raise her children in a loving, safe, fun, music-filled, creativity-glutted, God-celebrating environment.

Decades of her heart's desires.

She thought she'd been doing better. But maybe it was time to talk to Dr. Hanley about medication. She couldn't shake the two-taloned despair monster. The first of maybe many generations of Willowcrest children were being denied a music education, or at least an introduction to it. And she had been denied the opportunity for... for...

For more of what she loved. Did that define her as a woman with an insatiable appetite for her career? Or a woman with an unstoppable passion? Were either of them honorable enough for her to expect God to help bring about a conclusion she could live with?

She should be grateful for what had been and forget this nonsense that she could keep teaching—in a classroom setting, in addition to one-on-one—until she had no breath in her.

Bernie watched as Lucy counted out her tips and got ready to head home. He had no way of knowing her heart had been anywhere but at the restaurant that night.

She'd done what she needed to do, what was expected of her. And, for the most part, with a pleasant attitude. She lost a moment of cheerfulness with the catsup incident. She'd continue to not only put in her hours, but do a good job.

Every day she worked filling in for others, she was paid in tips and soul misery.

The second game of the double-header was still going when she arrived home. Charlie paused the game and asked how the night had gone. She told him. He laughed about the liver, which helped. A little.

She couldn't think of anything else to do, so she went to bed, mentally adding a new lie to her list. *If I were emotionally healthy, I wouldn't care so much.*

The problem was, the lies were starting to sound too much like truths.

"Is the game finally over?" Lucy sat up when Charlie came into the bedroom. She needed some ibuprofen or something anyway.

"Lucy. Oh, LucyMyLight." He shoved himself next to her and wrapped his arms around her. She could feel him trembling.

"What is it? Charlie?"

"Taylor Prentice."

"What about him?" Lucy sat straighter and slid over so Charlie had more room.

"The late news came on after the game was over. He drowned today."

Lucy fought to keep from throwing up. "What? No!"

"He and a bunch of other seventh grade boys were messing around on a raft out there by the resort. Birthday party for one of them, I think the newscaster said. Taylor got pinned under the raft. Caught in one of the anchor ropes. One of the kids pulled him out, but it was too late."

"Which friend?" Numbness. Shock. Stinging pain.

"I don't remember. Why?"

Dear God, how awful. How horrible to watch your friend die! How awful for Taylor's family. His little sisters. Lucy had them all in her classes.

The pounding in her head intensified. "Charlie, *when* did it happen?"

"You'll have to listen to the news report tomorrow. I think they said... one thirty. Something like that. I don't know."

Lucy shot out of bed and padded across the hall to her office. She booted up her laptop, the light from the hallway the only one she dared turn on, considering the pall that had settled over the whole world. A child was gone.

"What are you doing?" Charlie hung over her, his hands on her shoulders, steadying either himself or her.

"I have to know."

"Know what?"

Lucy scrolled through a series of calendar files, the ones she would have used this summer. Prepped months before the RIF letter. "There." She held back a scream with both hands.

"Lucy?"

She turned and buried her face in Charlie's chest. He held her as if she were the grieving mom. She should tell him, but couldn't.

"Can you talk to me?"

"I can't bear it!"

"I know all those kids meant a lot to you."

She shrugged out of his embrace and pointed an accusatory finger at her laptop screen. "Those boys would have been in summer band lessons at 1:30 today if the board hadn't—"

"Lucy, it could have happened any time."

"But it didn't. It happened then. And it wouldn't have. It wouldn't have. And he'd still be alive, and his friend wouldn't be traumatized for life. I can't... I can't bear this."

She knew her insides were flying apart, saw it happening in her mind's eye. But her hands were greased with more grief than she could carry, so she couldn't even catch one piece.

She stumbled across the hall and fell face first onto the bed. "I loved those kids."

"What?"

Lucy turned her face to unmuffle the sound. "I loved those kids. Every one of them I taught. The ones I could reach. The ones I couldn't reach. The ones that made my life light up, like Taylor and his buddies. Charlie, I can't bear it. I can't."

He sat beside her, rubbing her back, until she asked him to stop.

She was going to have to find a way to stop caring or she'd never survive.

At the funeral three days later, Lucy directed a small ensemble of Taylor's friends from middle school. They sang "Amazing Grace" and—at the parents' request—"All Is Well"—a song Taylor had sung as a solo at the previous year's Christmas concert.

Lucy's hands shook through the entire performance. But the young people did her proud. They held it together until the songs were over.

The honorary pallbearers were children. That should never happen.

Lucy's throat closed up completely when she greeted Taylor's parents and sisters in the receiving line. Taylor's mom reached out and drew her into a tight embrace.

"Oh, thank you so much for coming. You were Taylor's favorite teacher. He adored you."

Lucy's reply—"And I adored him"—squeaked out before she lost all ability to speak.

Charlie put an arm around her while she cried and said to each of the family members in turn, "So sorry for your loss. So sorry for your loss…"

Taylor's sisters clung to Lucy. Second and fourth grade. The line backed up behind the knot of them. Lucy didn't care.

Lucy remembered that shortly after her father died, one of the kids from his honors band died in a car accident. At the time, Lucy thought, "I'm glad Dad wasn't alive when this happened. It would have killed him." Crazy logic. But he might not have recovered from that pain. That's how deep his love for those children ran.

Hers ran parallel to his.

19

Another shift. Another few dollars. Another few traumas, like running out of the special of the day or spilling hot fudge on the rubber mat by the ice cream freezer or realizing that grieving the loss of a student couldn't easily be put on hold. Lucy took a moment to pray for Taylor's family before she pulled open the service entrance to Bernie's.

Lucy tied the black half-moon apron around her waist, then slipped her order pad into the pocket. Most of the other wait staff kept their aprons on hooks near the employee entrance and wore them a week or more before getting them laundered. Lucy washed hers after every shift. The smell of two-day-old deep fryer grease worked against her attempt at maintaining dignity. She couldn't give this role less than her best. The customers deserved it. Bernie deserved it. "Whatever you do in word or deed..."

Her powder blue polo was clean. Clean black slacks. Clean apron. Ready for her shift. She mentally rehearsed the past week's entry in the book club journal. The members had challenged each other to apply a familiar passage of Scripture to their relationships, especially with their HHATT husbands. "Create in me a clean heart, O God, and renew a right spirit within me." She'd copied it again, lower on the page, when she read it in the Bible she used at home. "Create a clean heart for me, God; put a new, faithful spirit deep inside me!"

Clean apron, clean hands, clean slacks, clean shirt. Clean heart would complete the picture. She had no problem serving. No problem meeting their needs. No issue at all with doing what she could to bring a bright spot to their day. On most days.

But it wasn't her heart's desire. She glanced at the blue tarp separating the restaurant from an unborn idea—Bernie's, not her own.

She wasn't sure God ever intended to let her have her heart's desire again.

"Lucy. Table Six."

"Got it." She drew a deep breath, dropped two pens into her apron pocket, and headed for the table in the corner. The customer sat with his back to her as she approached. Lucy recognized those familiar shoulders, that ruffle-able cap of hair with the slightest hint of red in the tangle of brown. "Sam?"

He turned, then stood.

She hugged him embarrassingly tight. "What are you doing here in the middle of the week?"

He hugged back. A flirt of fear coursed through Lucy's chest. He'd hugged as if something were wrong and he needed his mama.

"Sam?"

"Can we talk?"

What now? It was a little early in the shift for the dramas to start. Lucy looked toward Bernie, who'd observed it all, judging from the quirky smile on his face. He waved his hand as if to say, "Go on."

She slid into the low booth on the opposite side of the table. "Does Dad know you're in town?"

"He's on his way over."

The flirt of fear became a fist. "What's going on?"

Sam nodded toward the kitchen door. "Bernie said we could talk for a while as long as it doesn't get busy in here."

"This time of day? Not likely." She hated to nag. But she would. "So, tell me. What's up?"

Song of Silence

He drummed his fingers on the edge of the table. "If we could just wait for Dad..."

"This isn't something we can talk about at home? I'm not working long hours today. Are you in town for the whole day?"

"Dad's pulling in now." Sam pointed out the window.

Lucy looked in Bernie's direction again. He nodded and shooed her to stay at Sam's table. She sat and slid closer to the window to allow room for Charlie.

"Hey, son. How's it going?" Charlie gave SamWise a man-hug and patted him on the back, then slipped in beside Lucy. He'd been mowing the lawn. Grass clippings clung to his cap and grass juice stained the outer rim of his deck shoes. He smelled like the outdoors, in a fresh-air way.

Bernie set three coffee mugs and a carafe on the table. "Take your time, guys."

Lucy tried to gauge what she saw in her son's eyes. Remorse? No. Uncertainty?

It seemed like a fatherly thing for Charlie to get the conversation started with a well-timed, "Out with it, young man." Charlie instead said, "Boy, Martin and I killed the blue gills last Saturday. You should have been there. Had a nibble every cast."

Before Charlie could launch into the reason he now wore latex gloves when baiting his hook, Lucy intervened. "Sam, it's great to see you. Anytime. But you're here today for a purpose. You wouldn't have driven ninety miles for a cup of coffee with us. Something...serious, I'm guessing."

Sam looked off into space. He remained blank-faced a moment, then his brow furrowed. "Mom, where's the coffee house performance stage you were talking about?"

He was as distractible as his father. "First a permit issue. Now a slight miscalculation in the funding timetable. On hold for now."

"So, you're just—?"

Her raised eyebrows were enough to change the course of his conversation. "Okay. Here's what's happening." He crossed his forearms on the table. "I've been dating someone."

"We wondered when you'd get around to mentioning that." So that was the look. Her son was in love. "Tell us about her."

"It's about time you found someone." Charlie looked the part of proud papa.

"What's she like? Where did you meet her?"

Sam rubbed his triceps. "At church."

Charlie looked at Lucy. "That's a good start, heh?"

Could the restaurant get any quieter? Even the normal kitchen noise that filtered out into the seating area seemed muffled.

Sam stared into his cup and swirled the dark brew.

"Well, son," Charlie said, "thanks for letting us know about your girl. That's great. Isn't it, Lucy?"

"We're getting married Saturday."

The percussion of his news deafened her for a moment. She heard only red and white blood cells chasing each other through the veins in her head.

"Are we doing anything Saturday, Lucy?"

Dr. Hanley would not approve of the husband-threatening thought she briefly entertained. "Sam, there's more to this story. You're getting married this weekend?" She refused to blink until her son started making sense.

"She's not"—he leaned closer—"pregnant or anything like that."

Lucy shook her head, a bobble head of disbelief. "Still not...completely...comforted. Why such a hurry? We don't even know her name yet! We don't even know her name." She drilled the last words into the space between them. "You've always had a good head on your shoulders. I don't understand. We"—she glanced at Charlie—"don't understand."

"And...that may take a while." Sam rubbed his palms together.

"Are you looking for our blessing, Sam?" Lucy scrambled for what to say, how to get him to open up and tell the whole truth.

"Well, yeah. That means more to me than anything. Almost anything."

148

The table made more noise than the voices through the next several heartbeats.

"Her name's Sasha and she has a four-year-old son. Evan." His eyes softened. "So, as of Saturday, you'll be in-laws and grandparents."

"She's been married before?" Lucy asked, her tone matching the look in her son's eyes. Grandparents? This was now officially The Summer of the Unexpected.

"No. Never married."

Unexpected times forty.

"Is she pretty?"

Lucy poked her husband's arm. "Charlie? What on earth?"

"Just making conversation."

Sam laughed then. His beautiful, sweet, baritone laugh that had warmed Lucy's heart even during the colicky months. And now she was days away from becoming a grandmother. Step-grandmother. "Sam, could we start at the beginning?"

"Okay." He sighed again. "Yes, Dad, she's beautiful. I haven't known her all that long. But long enough."

Lucy had been thinking of switching to decaf. Today was not the day. "You haven't known her all that long? How long?"

"A couple of months."

Charlie scratched his head. "Your mom and I dated for eight years before I proposed."

"Dad, you met in middle school."

"Still..."

Lucy glanced up. The restaurant door opened. Bernie again waved her off and greeted the customers—another of Lucy's former students and his—

His grandmother.

Lucy took Sam's nearest hand in her two. "You knew the suddenness of this would rattle us, Sam. Tell us what's really going on. Love happens suddenly sometimes. We understand that. A woman's past doesn't have to chain her to a hopeless future. A child is a child. It's what you're not telling us that's making you avoid eye contact and making it hard for us to say, 'Congratulations!'"

"Congratulations!" Charlie said.

Lucy bit her bottom lip.

Even in adulthood, Sam's sweet face could charm moss off a rock. To see it wrestling with what should have been celebratory news skewered Lucy. She squeezed his hand. "Sam, whatever it is, we're here for you. We love you. We trust that you've talked to God about this." *Charlie, step in here any time.* "We want the best for you, honey. If you love her—Sasha—we'll love her, too. And her son."

"Steven," Charlie leaned in to whisper.

"Evan," she whispered back.

SamWise exhaled noisily. "He's a great kid. He just—"

"Just what?"

"He and Sasha haven't had the easiest life. But she's an amazing woman. I know we're doing the right thing."

"And you love her," Lucy added.

Sam drew his hand back. "Love looks so different than I thought it would."

Olivia, I wish you were here right now. And I wish you'd already finished your masters in psychology. Or doctorate. Even better.

Lucy excused herself long enough to beg the rest of the day off. Family crisis, she called it, regretting the word *crisis* as soon as it left her lips. Bernie was Bernie. Accommodating. Understanding. The perfect boss. He called his wife to fill in for Lucy and sent the trio on their way with his blessing.

The three had driven separately, so the ride back to the house—which is where the conversation should have taken place originally, in Lucy's opinion—consisted of three cars, three drivers, and an accompanying battalion of uncertain thoughts.

"Want a sandwich, Sam?" Charlie asked when they'd settled at the kitchen table. "I can make grilled cheese. And not just any grilled cheese."

"Not really hungry, Dad. But thanks."

"Can I make one for you, Lucy?"

"I'm not hungry either."

"Mind if I...?"

"Go ahead," Lucy and Sam said in unison.

Sam turned toward Lucy. "Mom, I love Sasha and Evan. But... it's complicated. Relationships can be complicated."

Sam's father whistled show tunes while toasting bread and slicing cheese twelve feet from where Lucy and Sam waded through the next phase of conversation. Yes, relationships could be complicated.

"I know this isn't how you'd envisioned my finding the love of my life. Not the lead-time you thought you'd have for your son's wedding. But it's going to be okay."

"How do you know that?"

"Because I'm committed to making this work."

"Sam, marriage is hard enough when you're deliriously in love with one another."

"I heard that," Charlie said.

Lucy took a slow breath. "Can you two take some time to think about it? At least that?"

"Mom, they need me."

Lucy had expected to be strong, decisive, persuasive, caring, but logical. She hadn't anticipated the tears. "Whatever their problems, honey, you can't be their savior. It never works when a human being who isn't God tries to fill that role." Good grief! Where were these words coming from? They almost sounded like... wisdom.

Head bowed, Sam said, "I can carry on what He started."

Well. She'd managed to raise a selfless son. Well. What could she say to that?

"Does Sasha love *you*?"

Sam's smile returned. "I didn't envision ever having a discussion like this with my mother."

"Well," Lucy said again, "your dad's making grilled cheese, so..."

That line broke the tension. Sam left his side of the table and circled around to hug his mom. His arms felt so good around her, proof that this insane decision that he couldn't even explain hadn't come from a place of rebellion against all he knew and all they were as a family. It was something more, deeper, something that defied anyone's ability to put into words. Rooted in love, but a different kind of love than she was used to or had seen from her son before.

"Marriage is a serious matter, SamWise," she whispered.

"I know, Mom. I've watched yours." His embrace tightened around her. "And I've seen friends marry for what they thought was love but turned out to be infatuation. And even dumber reasons than that." He took the chair next to her this time. "I'm committing myself to Sasha and Evan because I want to join my heart and life to theirs. She's a beautiful spirit. But she hasn't had it easy. I want to give them a family. And that includes you and Dad and Olivia..."

Charlie set his sandwich plate on the table. "Tell her the woman is into music, and your mother will be all over this wedding idea, son."

Sam reached over and snagged a small corner of the grilled cheese. He held it between his fingers, staring at the crumb like a person might contemplate a communion wafer. "Music? She loves it. Not in the traditional sense," he said. "She's completely deaf."

20

In the past, Lucy had stopped just short of pride that she entertained few true fears. She managed to eliminate or relocate spiders on her own. She faced off against the harmless snakes that infrequently took up residence in the garden. If they'd lived in a part of the country where venomous snakes were the norm, she might have had a slightly different attitude. Lucy plowed through storm warnings and would have welcomed roller coaster rides if Charlie hadn't had an unfortunate experience as a child that turned him off of amusement parks.

Her list of fears barely constituted a list. Fear of her children making decisions that brought lifelong regrets. Fear of losing her ability to sing. An unhearing life.

Two of her three loom-large-destroy-life-as-we-know-it threats were embedded in the news that Sam and Sasha already had their marriage license in hand and planned to make it official in their church chapel on Saturday afternoon. Potential for lifelong regrets from a rushed wedding coupled with a soundless existence. Lucy could not imagine the situation being any more traumatic. If it were her. But it was her son's life. His to choose. His puzzle. Somehow, she was going to have to add it to her pile of adjustments and find where the hope was hidden in the smoldering pile.

"Is Sam here? That's his car, isn't it?" Olivia's questions started as soon as the front door clicked open. "Hey, Sam!"

He'd stood when Olivia walked into the kitchen, always the gentleman Lucy had—*they* had—taught him to be. Olivia bear hugged him, then asked a question with so many interpretations. "What are you doing here?"

Filling her in took less time than it had taken Lucy and Charlie to grasp the reason for his visit.

"Dude!" Olivia's one word initial response. It seemed so out of character for her that Lucy had to stifle laughter. No laughing matter. No.

"What can we do to help?" Olivia's follow-up response.

Why hadn't Lucy thought of that? She'd been mired in concern. Worry got in the way of a lot of good sentences. "Do you want us there on Saturday? Do we need to pull together a rehearsal dinner?"

"It's a very low-key ceremony, if you can call it that," Sam said. "We won't need a rehearsal or rehearsal dinner. No family on Sasha's side."

"None?" Lucy couldn't imagine that. How alone the young woman must feel.

"None here in the United States."

"A few more stories yet to tell?"

"Eventually," Sam said.

"So, how can we help? Really?" Olivia—relentless in all the right ways.

"We need someone to help with Evan."

Olivia scrolled through her phone. "I don't work until five that night. I can watch him for you. You're talking a short service, right? He's four. It'll be fun."

"He doesn't connect well with people."

"Only-child syndrome, huh?" Olivia pulled the band from her ponytail and reinstalled it.

"How far have you gotten in your psychology pursuit, sister of mine?"

"Comes naturally to some of us," she said, fanning herself and tilting her nose in the air.

"Good," Sam said. "You'll need it. And you'll get an afternoon of hands-on—or I should say hands-*off*—experience with Asperger's."

Normally Lucy wouldn't notice when the air conditioner cycle kicked in. But when the room is that silent...

"Sam, you're sure you can do this?" Charlie laid his arm across Lucy's shoulders.

"Completely sure I can't," he said. "Not without your help, God's help, and some heavy-duty prayer."

"Already started," Charlie said.

Lucy laid her hand over his. She drew strength from the texture and weight of that hand, even the roughness, a sign of rugged faithfulness she'd counted on since high school.

"Now, the last thing." Sam held his temples with the heels of his palms.

"There's more?" Lucy's pulse staccatoed.

"Oh, two things. One, we're rushing this to Saturday because the family Sasha's been staying with is moving to South Carolina. We thought we had a little time because they had to sell their house first. The one time you wish a house would have stayed on the market longer..."

"So, they're moving in with you," Olivia said. "Hey, I just realized you two saved me a bank vault of money on a bridesmaid dress. Thanks, brother."

"They can't move in with me. The apartment complex has a *no kids* rule. That's what appealed to me about it. Ironic, isn't it?"

Lucy glanced toward the basement door. "Ironic."

"Sasha's an architect for a company in Hudson. She can work remotely from anywhere. Technically, I can too, if I'm okay with traveling from time to time. Which I would be if I knew Evan and Sasha were..."

"Are you asking if you three can move in here?" Lucy worked hard to keep her voice from betraying the anxiety that clogged her airways.

"Whoa," Charlie said.

"No. No, guys. We need a place of our own. I'm not sure how quickly we'll be able to find one. That's all. And wherever we land, we have to have skilled daycare close by. Sasha doesn't have a driver's license. Yet. When I'm on the road, that could be an issue."

"Son, it sounds as if your life is about to become one big old box of issues. Huh. Box of issues. Tissues." Charlie seemed especially pleased with himself.

Olivia shook her head so Lucy didn't have to.

"Are you looking for a place to stay temporarily? Until you find something near us? Near home?" Lucy hadn't expected to be a revolving door parent, with grown kids moving in and out and in again.

"I...I don't know. I can't make my problems your problems. That wouldn't be fair. But, yes. Temporarily. Sasha and I have a lot to work out. I'm not even sure she'd go along with the idea of our camping out here for a while. She's very...private. A lot of reasons for that."

"While it's still summer," Charlie said, "I don't mind sleeping in the basement. Can you three make do with the master bedroom?"

"Charlie!"

"What?"

"You may not mind sleeping in a store room, but I should probably be consulted, don't you think?" Lucy clenched her fists under the table. She released their tension and flattened her hands on her thighs.

Olivia stood. "How about if I make up some lemonade for us?"

"Great idea," Lucy said, her teeth motionless as she spoke. "Charlie, you're talking about putting two newlyweds and a four-year-old in the same room?"

"We could—"

"What is it, Sam?"

"We could take the basement. In our off hours, we could work on doing some of the renovations you always talked about."

"That'd be fun," Charlie said. "I love a good project."

The house wasn't done shrinking. Three more people. Temporarily. Even Olivia's presence was temporary. The new family might find something quickly. They might.

"It's better in a lot of ways if you two aren't relegated to the basement," Sam said, his eyes as apologetic as Lucy had seen them since the permanent marker on the couch incident years ago. "The way we get Sasha's attention if she's not looking our direction is stomping. She can feel the vibration through her feet and then knows we want to talk to her."

"She hears nothing?"

"Nothing. Mom, don't pity her. She's a very capable person. One of the strongest women I know." He looked from Lucy to Olivia. "And I've known a couple of incredibly strong women."

Olivia stirred ice into the pitcher. "He meant that as a compliment, right?"

"I'm taking it that way," Lucy said.

"Would it help if I moved out?" Olivia asked. She poured four glasses of lemonade.

Would it help if I did?

Swah. Sweh. The one thing Lucy noticed was the sound her thumbnail made scraping the fabric of the chair arm in Dr. Hanley's office. A low *swah* one direction and *sweh* the other. The A below middle C followed by the D above it. And breathable silence between.

"We don't have to talk at all, if that's what you want, Lucy." Dr. Hanley's voice had the texture of custard. Rich. Not too sweet.

"I don't mind talking it through." *Swah. Sweh.*

"You have a potentially wild ride on your hands. Didn't you tell me once you were *into* roller coasters?"

"I was younger then."

Swah.

Dr. Hanley crossed her legs at the ankles, bronze sandals revealing well-manicured feet. "May I ask you a question?"

"I'm running low on answers."

"Would you describe the series of events since May as individual bricks or as pop-beads, strung together?"

Lucy listened to the silence a moment before responding. "Pop-beads, I guess. No space between them. Linked to each other."

"That's natural. But what if they were separated? Taken individually?"

"That's not how it happened."

"Humor me?"

Lucy would likely have to give up her sessions with Dr. Hanley. A basement remodeling project wouldn't happen without money to buy the materials, despite Sam's offer to pay for the lumber and drywall. She'd better humor her counselor.

"What if"—Dr. Hanley began—"your son had simply announced he was getting married. That's all. How would you have responded?"

"I would have been happy for him. He'll make a great husband."

"What if all you heard this week was the news that you're about to become a grandparent?"

"I'd be ecstatic."

Dr. Hanley smiled, then brought her expression back to neutral. "Taken all by itself, how would you have responded if you heard that this grandson coming into your life was not a newborn but four years old, that your son was basically adopting him?"

"I'd be proud of my son and eager to meet my grandson."

"How would you and Charlie have responded if Sam had met you at Bernie's and asked only one thing—if he could move home for a while?"

Lucy raised her hands in surrender. "I get it. Taken individually, none of the announcements seem as threatening as they do when lumped together."

"Or in our current analogy, linked together like pop-beads. Part of your reaction to Sam's story started back in May. You began to expect disappointment. When another one showed up, it added to the gravitational pull of the one before. Then Sam arrives and pop-pop-pop-pop-pop, the string of disappointments is so long it gets tangled in your arms and feet. It wraps itself around your throat. It gets in the way."

Swah. Sweh. "A person can choke on a single pop bead."

"Granted. No metaphor is perfect."

"You know what I find interesting?" Lucy cupped her knees with her hands. "I'm not heavy-hearted over what this will cost Charlie and me. Except for one thing."

"The threat to your solitude?"

"Okay, two things."

Dr. Hanley tapped her pencil eraser against her cheek. "Your loss of freedom to come and go as you want or need to?"

"Okay, so more than a couple of things. The point I wanted to make is that I'm afraid my Sam is getting himself into..."

"Into what?"

"Oh."

"Into pouring himself out for the needs of others? Like you always have? Into seeing a need and wanting to meet it? Into building a marriage out of mutual respect rather than merely physical desire?"

"Into his life's passion."

Dr. Hanley held the box of tissues toward Lucy. "It doesn't look anything like you thought it might."

"No."

"Sometimes a calling surprises us."

"Sasha and her son aren't charity cases." Lucy said it as a reminder to herself as much as a statement for Dr. Hanley's ears.

"No. The minute any of us starts thinking that way, relationships suffer. And God is likely to rattle you with one of Sasha's or Evan's strengths that far surpasses your own. Just to prove His point." She tapped her chest with her fingertips. "Ask me how I know."

Lucy's breath patterns had changed since she'd arrived for the appointment. If she only had a room like this she could escape into at home. Or—

Or a virtual room in her soul.

"We haven't worked out the financial arrangements yet," Lucy said, "or what the schedule at home is going to look like. So I'm not sure if I'll be able to keep booking appointments with you."

"You may need someone to talk to all the more, with what's going on. I'm not trying to solicit business for my practice, you understand."

"Assuming I can find a way, what's my homework assignment for this week?"

"From the moment you meet Sasha and Evan, see them as gifts to your son, your family. Many gifts come with responsibilities, but try to maintain that sense of discovery and delight as those gifts are unwrapped."

"We're working like mad to get the basement rearranged enough so that they have a place to sleep. We can't get the remodeling done with this little notice. So it'll be a mess for a while."

"Families always are."

"Point taken."

"And we still have to choose a wedding present. Should it be practical? Or romantic?"

"This." Dr. Hanley moved first one hand then both.

"What is that? Sign language?"

"American Sign Language. Yes. First make a W with one hand. Pinch your thumb and pinky together. There. W. Now touch the index finger of the W to your mouth and move it outward."

"Like this?"

"Now, using both hands, bring your thumb and index finger together with your other fingers extended. Start with your thumbs touching, here, close to your heart, and swing your hands in sweeping arcs until your pinkie fingers meet on the far side of the circle."

The movement felt awkward. Lucy tried again. A third time.

"Good. Put the two actions together. The extended W followed by the encompassing circle."

"What does it mean?"

"They're your first words to Sasha. 'Welcome to the family.'"

21

Lucy and Charlie had visited Sam's place of worship a handful of times in the years since Sam had started attending the not-quite-mega church. Concerts, comedians, theatrical productions, special events. The dichotomy of the primarily contemporary service held in a building with ornate carvings, antique stained glass windows modernized and protected with new framing, and the massive pipe organ always impressed her. The best of the elements of adoration and majesty from a heart of untethered, intimate worship. If it hadn't been a ninety-mile drive each way, Charlie and Lucy might have made this their church home.

In all their visits, they'd never had reason to see the chapel Sam and Sasha had reserved for their wedding.

Wedding. The word still sat awkwardly in Lucy's mind, as if perched on a nerve ending that swayed with gusts of wind.

The room would have read monastic in its simplicity if not for the modern music piped in, the small raised platform at one end, and the curved floor-to-ceiling paned windows that offered the chapel a panoramic view of a courtyard rose garden.

Olivia nudged her mother. "Wow, Mom. Isn't this gorgeous? Ideal. I want to get a picture of the three of us with that in the background." She pulled out her phone. "And silence this while I still have a chance."

The bride wouldn't be disturbed by phones chiming in the middle of the service. How differently Lucy thought about sound since Sam's announcement. The soft ambient music in the background set a mood—serene and worshipful. It was as much a part of the décor of the chapel as the view of the rose garden. Sasha wouldn't hear it.

Olivia waved for her parents to join her near the windows. A side door opened and let in a small stream of people. Olivia hustled off the platform and joined her parents in the middle of the short center aisle.

Sam's smile lent a new light to the room. He stepped around the associate pastor and his wife—Lucy had met them before but couldn't pull up their names—and hugged his family. "It's so great to have you here to witness this." The glistening in his eyes was unmistakable. "Dad, Mom, Olivia, this is Sasha and her son, Evan."

Sasha had the figure and carriage of a ballerina or model, with the exception of the four-year-old propped on her hip, his legs whipping through the cloud of fabric of her elegant but dramatically simple ivory ankle-length dress as if he were frantically treading water. Sasha slid her hand down his leg and stilled it against her side.

"Sasha, my father, Charlie." The pastor's wife signed the words as Sam spoke. Sasha watched her, then looked at Charlie.

Charlie nodded and smiled.

"My sister, Olivia."

Olivia found a way to hug both Sasha and Evan at the same time. Evan squealed, but quieted as soon as Olivia backed off.

"And my mother, Lucy." Someday Lucy would ask why the signed word for mother touched the chin, not the heart.

Extended W. Encompassing circle. *Welcome to the family.*

It wasn't her new daughter-in-law but her son who responded. Thumb extended, he touched the fingertips of his flat palm to his strong chin then pointed the fingertips toward her. His *thanks*.

Sasha stepped forward. She lowered Evan to stand as she held Lucy's gaze. Her hands moved in an elegant, brisk, allegro conversation Lucy couldn't understand. The interpreter behind Sasha tapped her on the shoulder. The movement stopped.

"Please tell her," Lucy said to the interpreter, "I'm so sorry, but 'Welcome to the family' is all I know."

Sasha's lip-reading skills must have been honed early in life. Before the interpreter signed Lucy's words, Sasha signed a response then watched Lucy's face.

The pastor's wife gave voice to her message. "'That's all I needed.'" Sasha tilted her head, smiled, and signed "Thank you."

Such a beautiful smile. Riveting dark brown eyes, but rimmed with a pain yet to unfold.

Evan—his eyes as bright blue as hers were dark—wouldn't let Olivia near him until she pulled out her smartphone and earbuds. She glanced up at Sasha and her brother. "Is it okay? I downloaded a drawing app that has virtual stencils to trace."

Sasha looked at the screen Olivia held out to her, then nodded.

"Come on, Evan. Let's sit here in the second row and have some quiet fun."

Sasha signed a short sentence.

"What did she say?" Lucy asked the interpreter.

"'Good luck with that.'"

Olivia glanced up, her almost impish facial expression showing how ready she was for the challenge.

Pastor Davidson directed Lucy and Charlie to sit in the second row on the other side of the center aisle, opposite Olivia and Evan. Balancing the small audience? Sasha and Sam sat across the aisle from one another in the front row. The interpreter—what *was* her name?—stood to the pastor's right on the raised platform, since the traditional bride's side was on the left facing the front. Although few things about this wedding seemed traditional so far.

The pastor nodded toward the back of the room. Lucy turned. A videographer had taken his position. Pastor Davidson then opened his Bible, positioned a page of notes on the open book, and said, "It's time to begin a new phase of life for these two sojourners. We'll start where all hope-filled journeys begin—with prayer."

When his prayer ended, the ambient music swelled. From her perspective, Lucy could see the bride lift the hem of her skirt and kick off her jeweled sandals. She pressed her bare feet against the hardwood floor, and soon began to sway in perfect rhythm to the flowing music. She must have been sensing the vibrations in the soles of her feet.

Sam bent down and removed his shoes, too.

As the song reprised a familiar passage, Sasha stretched her arm and leaned across the aisle, her head tilted so her ear almost touched her shoulder. The aisle was too wide. In time with the song, she withdrew her arm and put her hands in her lap. Sam stretched his arm across the aisle, leaning in as the music dictated. Even his arm was too short.

Pastor Davidson took one step forward and spoke above the music. "The aisle will always be too wide for a husband or wife to bridge the gap alone in marriage."

Slowly, riding the waves of the music, Sasha and Sam stretched their hands across the aisle, leaning toward the other. When their fingers touched, she slid her hand into his. They stood, stepped toward one another, then onto the waiting platform. Barefoot. Barefoot in church. It had never seemed more appropriate.

Sasha's hair—darker and curlier than anyone's in the Tuttle family line—was caught up in an almost Edwardian style, with tendrils floating around her shoulders.

"The couple has chosen a familiar passage of Scripture," Pastor Davidson said, "which they'll share with you now."

Sasha and Sam turned so they faced the four seated in the chapel. And the videographer. Lucy tried to imagine not having the gift of hearing. Would she have recognized the verses without sound, only from the signing the couple did? Clanging cymbals. Sounding brass. Noise. Yes. The opening verses of 1 Corinthians 13. So familiar, Lucy often tuned it out at weddings.

Not today. She heard and saw the message expressed in the hands of her son and his bride. Sasha watched Sam, often waiting for him to catch up, or reaching to close his fist or stretch his fingers to an open position to more accurately reflect the words' intent.

"Love is patient," Pastor Davidson read. Sam and Sasha touched their thumbnails to their chins and drew them down about two inches.

"Love is kind." The bride and groom laid an open hand on their heart with the middle finger touching their chest and rubbed lightly in a circle.

The pastor read slowly, pausing as each attribute of godly love found its sign language counterpart.

"Love"—crossed arms, hands closed as if grasping—"trusts, hopes, endures."

"Now faith, hope, and love remain—these three things," Pastor Davidson read. "And the greatest of these is—" He stopped and extended his palm toward the witnesses, indicating that they would communicate the last word.

Lucy caught movement out of the corner of her eye. Evan didn't look up, but handed the phone to Olivia and crossed both hands over the middle of his small chest. The adults followed his lead.

Pastor Davidson shared a short charge to the couple, which always wound up being a reminder to the married couples in attendance—Lucy and Charlie—what to return to, and to the singles in attendance—Olivia—what to aspire to.

Lucy watched the interpreter's gestures grow more dramatic when the pastor raised his voice for emphasis. The signing world was one about which Lucy knew so little. Crash course on the way.

Sam and Sasha had written simple vows to each other. Poetry in the simplicity and sparseness of words. Lucy couldn't help wondering if she'd missed something in trying too hard to express her love in words over the years. Fewer but richer words. Would that have worked?

They exchanged their vows in a way beautifully fitting to their unique situation.

"I give you the promise of my forever," Sam said.

She read his lips and replied in sign, "And mine...to the Keeper of our forevers." The interpreter, her face intentionally neutral when not signing, wiped a tear after signing *forevers*.

"I give you my promise of safety in my embrace." Sam's strong chin quivered. So many stories yet untold.

"I give you my faithfulness without hesitation." The interpreter's voice, but Sasha's heart.

"I will not treat your wounds as if they are nothing," he said.

"And I will not let my wounds scar our tomorrows," she signed.

Sam led Sasha by the hand to a low wooden bench near the windows. She tucked her flowing skirts to the side as she sat. Sam knelt and drew a basin and towel from under the bench. Lucy glanced at her daughter, who had one arm on the back of Evan's chair and her other hand pressed tight against her mouth, tears streaming down her cheeks. Like mother, like daughter.

Sam dipped an end of the towel in the water and, like Jesus had done with His disciples on His last night with them, washed Sasha's feet in an act of unexplainable, indefinable love. He drew a small pottery vase from under the bench and poured a small puddle of whatever it contained into the palm of his hand. Oil? Perfume? Lucy smelled nothing but a faint scent of pomegranates. He smoothed the oil onto the tops of her feet and ankles.

Lucy almost looked away, the scene so intimate.

He wiped his hands on the towel and slid the basin to the side. Then he took Sasha's hand and helped her stand. He tried to direct her to the center of the platform, but she tugged him toward the bench and pressed on his shoulders until he sat.

Sam shook his head no. Sasha hiked up her skirts, knelt at his feet, and drew the basin of water toward her. She repeated his actions, including the oil, then laid the towel in her lap and lifted his feet to rest there while she signed.

Sam looked to the pastor's wife, who'd slipped close enough to see what Sasha was saying. The interpreter fought back tears. It took her a moment before she spoke. "Sasha said, 'In the name of the Father, and the Son, and the Holy Spirit, my husband. May you know His grace as you've shown it to me.'"

Charlie leaned over to Lucy and whispered, "If that isn't love, I don't know what is."

Sasha was the only one in the chapel who hadn't heard what Charlie said.

Lucy turned to see if Olivia had managed to hold things together. She and Evan were gone. Which one of them had a meltdown?

The couple had no rings to exchange but intended to get them soon, the pastor explained. A few more words. Another prayer. A sweet, non-lingering kiss. A benediction of blessing. Cue the music. Lucy watched Sasha's face register when her feet picked up the vibration of "Ode to Joy."

22

The newlyweds wanted Evan in some of the wedding pictures. Lucy volunteered to find out where Olivia had taken him.

She found them both in the women's restroom. Evan stood in a corner, throwing one balled up paper towel after another in Olivia's direction. They landed a few inches from his feet every time, but that didn't stop him. The blank stare would only need to deepen a few degrees to become downright scary.

Olivia had the stainless steel waste can in her hands, using it as half-shield/half-receptacle, attempting to catch towels aimed at her.

Lucy sniffed. "Something smells ripe in here."

"Oh, Mom. It's Evan." She set the can on the floor and walked away with her hands in the surrender position.

"I...assumed it was him." Lucy tamped down her comedic response.

"Did you see us walk out?"

"No."

"Good. He just slid off the seat and headed for the back door. I didn't want to holler, so I followed him. I tried to pick him up so I could talk to him, but he won't let me touch him. At all. So I just walked behind him while he went exploring. He came in here, which I assumed meant he needed to go to the bathroom.

Well, how was that going to happen if he wouldn't go into one of the stalls?"

"Oh, honey."

"Once the...odor...became apparent, I thought I could at least get him out of his soiled stuff so we could salvage the dress pants. But no way. I told him I didn't have to do it, that he could take care of it. Nothing registered with him. He's four, right? Probably potty-trained for two years by now, wouldn't you think?"

"I'd better go get Sasha. Maybe she brought extra clothes for him."

"Bring the S.W.A.T. team, too. His throwing arm is getting better. I may need back up."

Lucy plopped her pink leather journal on the table and dug for her pen. "Okay, ladies. Your turn. How was *your* week?"

"Completely uneventful," Carole said.

"Blissful, even," Angeline added. "Compared to yours."

Marta patted Lucy's hand. "Not a one of the rest of us became unexpected mothers-in-law and grandmothers this past week. That honor's all yours."

"I wish the newlyweds could have had some time alone together. Even one night away somewhere. But Evan shouldn't be left alone with us—with people he considers strangers—right now. That's asking for trouble."

"How long before he'll get adjusted to your family?"

"Good question," Lucy said, popping a breath mint into her mouth. "*I'm* not completely adjusted to them, as you know, and I've lived with Charlie for decades."

The laughter felt healing after days of tension and uncertainty and a future that promised more of the same.

"It just goes to show you..." Angeline said. She picked at the edge of the napkin under her soft drink.

"Show us what?"

Angeline looked up. "I'll think of it in a minute."

Another round of laughter.

"Did you have a reception or anything?" Marta, the one with the gift of hospitality.

"They insisted they didn't need one."

"And you believed them?" Marta, again.

Carole nudged Marta's forearm. "What she means is that they probably didn't want a fuss made over them, but there are other ways a community can help a new couple. We could do a card shower for them."

Marta spoke up. "Lucy's the only one from this community."

Angeline opened her purse and removed a bill, which she slid across the table toward Lucy. "There. You tell them congratulations from me."

"Angeline! A hundred dollars?"

"Ooh! Was that a hundred? I thought it was a dollar."

Lucy held her breath.

"I'm kidding." Angeline clapped her hands in front of her face, obviously pleased with herself. "And maybe that's the thought of the week."

Carole opened her journal and uncapped her pen. "What is?"

"That sometimes you have to dig for the humor in it. Husbands who can't find the mayo in the refrigerator. Being grateful wedding photos don't have smell-o-vision. Letters like the one my widowed friend Lynda got yesterday, addressed to her dead husband, suggesting he had only one issue left before expiring."

"Oh, no!" Lucy nearly choked on her breath mint.

"Lynda said she wrote back and told them he'd expired already and for the first time in his life had no issues at all."

Carole giggled and stopped writing in her journal. "I had another thought picked out to share, but that's a good one. Dig for the humor."

Lucy's jaw hurt from where she'd bitten down on the breath mint. "What was it? The other thought?"

All eyes turned to Carole.

"I've been playing the martyr lately," she said. "For the last...thirty years or so."

Silence.

"If my husband asked me to stop at the bank for him, for instance, I'd sigh and say, 'I guess so.'" Carole doodled in her journal. "I knew good and well that I was going to do it for him anyway. I love him. Even if it did take a few more minutes out of my day or make me have to backtrack. Of course, I'd go to the bank, if that's what would bless him. Or help him. Or, frankly, just because he asked. Why did I have to sigh? Why say, 'I guess so,' as if it was a bigger imposition than offering him a kidney if he needed it? I was going to do it anyway. What purpose did it serve to let him know he was inconveniencing me?"

"Especially," Marta said, "since he may have heard the sigh but not connected it with the idea that he was inconveniencing you."

"True," Angelina said, nodding.

"Bad posture," Lucy said.

"My chiropractor's working on it." Marta straightened, grimacing.

"It's like bad posture when singing or playing a musical instrument. You're going to play the note anyway. Why not give yourself every advantage, every opportunity to make *beautiful* music, not a sour note due to lack of breath support or pinched off airway? You're...going...to play the note. Play it well."

"My mom played the martyr," Marta said. "I hear her in my voice sometimes when I feel compelled to make sure my husband understands what his request is costing me."

"Sometimes," Lucy said, the neck of her shirt suddenly constricting, "I think I have more than enough on my hands right now, thank you very much. I can't focus on my marriage relationship at the moment. If I'm short with Charlie, that's just part of the fallout of the crises we're juggling."

"But...?" Angeline's eyes widened.

"Oh, you knew that was coming, didn't you? But the stronger my connection with my husband, the better my ability to weather all the other storms."

Marta sat back. "And there you go."

Lucy's appreciation of this most unusual book club grew every time they met. A unique collection of personalities. Straight talk. Grace-filled talk. "I have to leave early today. We're meeting with Evan's new doctor in a few minutes so we understand what's going on in that little one's mind. As sweet as he can be, this is a round-the-clock challenge. For all of us."

She stood and gathered her HHATT journal, pen, and purse. "Oh, I need to pay for my tea."

"Got it," Carole said.

"You don't have to do that." Lucy rummaged in her purse. Carole stopped her with pursed lips and raised eyebrows. "Okay, okay. Thank you." Lucy bowed. "I'm grateful. I don't want to be late for this appointment. This family needs all the help we can get."

"The home situation you're describing," the pediatrician began, "is the worst possible environment for Evan right now."

From the slump of Sasha's shoulders, Lucy knew she'd lip-read every word accurately.

"The sensory overload of a new place to live, three complete strangers plus a new father figure, three of the five adults working from home..."

"Oh, I don't work," Charlie offered. "And when Olivia's home, she's not working. She's studying for a degree in psychology. Online. All online. It's the craziest thing."

Did he think that would help the cause?

The pediatrician nearly succeeded in suppressing his amusement. "I don't know what to tell you, folks. Are you prepared for the kind of chaos about to be unleashed in this little boy's system? Are you willing for him to pay that price?"

Sasha signed something to Sam. He shrugged. "I'm sorry, Sasha. I don't know what you said."

Brows furrowed, she cupped her hands, fingertips touching in front of her, then flicked her hands, palms down, fingers spread. And stomped her foot.

The non-verbal communication would take some getting used to. The musicality of what she said with her hands, with her whole body, in the absence of tone, vocal inflection, verbal dynamics, volume...

"No, it's not hopeless," the pediatrician said.

Sam seemed relieved on more than one level. The doctor knew what she was saying.

"But"—the doctor shook his head—"it's not good. Anything you can do to cut down on the overstimulation..."

"He did okay at the wedding," Sam said. "My sister gave him ear buds and her smartphone."

Lucy rehearsed the memory of the bathroom scene. Would she call that doing "okay"?

"You may need to rely on technology more heavily during the adjustment period," the doctor said. "He may need to tune out everything else that's happening in order to function at all. A single song. A single game or movie. One person. One of you."

"In addition to me?" Sam said.

The doctor removed his glasses. "Sam, you pose a double threat to Evan right now. You're part of this new dynamic, the move, these people you connect with but he can't. And...you've won the heart of his mother. He can't empathize with why you'd want to spend time with his mother, or why Sasha would want to spend time with you. You can't let that keep you from bonding as a couple, but he's likely to act out when he sees the two of you together. For a while."

Sasha seemed to catch every word. Lucy noticed that the pediatrician enunciated clearly, almost theatrically, no doubt for the young mother's sake. That's how Lucy should have described it to her students. She should have encouraged them to enunciate when singing as if trying to help a deaf person read their lips.

"I would encourage you," he said, "to find a place of your own just as soon as possible. Let's pray we can stay on top of this for Evan's sake, and for your sake, too."

They found Olivia and Evan in the clinic's large waiting room. Evan stood nose-to-the-glass of a massive aquarium.

"He's been like this almost the whole time," Olivia said. "I tried to get him to back away a couple of inches, but he is"—she put one hand on her blue-jeaned hip—"as stubborn as he is adorable."

Sasha mouthed and signed a sentence Sam interpreted for them as, "Tell me something I don't know." Then she thanked Olivia for watching her son.

"Evan, let's go," Sam said, his voice calm but firm. "Come on, Evan."

Sasha stomped twice. Evan turned.

"Does Evan have hearing problems, too?" Charlie asked.

Sasha signed an answer as Evan drew near. Sam snickered.

"What did she say?" Charlie cupped his hand behind his ear, as if that would help him hear her hand movements.

"'Yes. Hearing problems,' she said." Sam leaned toward his dad. "Nothing to do with his ears."

Charlie smiled. "That's boys for you."

It's so much more than that. So much more. But Charlie wasn't entirely wrong.

Sasha seemed to study the high ceiling in the waiting room. A series of skylights sent shafts of sunlight onto the atrium garden in the center of the room. Striking. Sasha must have thought so, too, the way she held Evan against her leg and let her eyes wander from corner to corner to corner, taking it all in.

Somehow the elegance that defined her on her wedding day lingered like a sweet fragrance woven in her hair. She glided toward the door, despite the child attached to her.

Lucy's thoughts ping-ponged from the words they'd heard in the clinic's conference room to the wordlessness of the fascinating woman a few steps in front of her. Getting to know her daughter-in-law better wouldn't be easy because of the...language barrier.

Lucy corrected herself. Sasha wasn't wordless. Her words had shapes, and emotion, and meaning. No wasted action. Everything meant something. She wasn't music-less either. The young woman picked up her son and waltzed him to the car. Evan lay his head on her shoulder, letting his mother lead.

23

"So now what?" Charlie said when they'd all piled into the family room.

"Dad, keep it down, okay?" Sam gestured toward Evan, asleep on his shoulder.

Sasha eased Evan out of Sam's arms and left the room, no doubt headed for their basement cave.

"Want to talk here or in the kitchen?" Lucy asked. She adjusted the thermostat to the "occupied" temp. Olivia excused herself to work on an assignment for one of her online classes.

Sam sank onto the couch. Charlie took his recliner—the least recliner-looking chair the furniture store had in stock—and Lucy sat in her reading chair.

"We need Sasha in on this conversation," Sam said.

"The girl looks like she needs a nap, too." Charlie slid back and popped up the footrest.

"What every woman dreams of hearing," Lucy said. Hearing. The most common sentence held new meaning now. How long would it be before she didn't second-guess everything she said in this house?

Sam leaned forward, hands clasped in front of his knees. "This was never intended to be a permanent solution." He dropped his chin to his chest. "I may have made a disastrous mistake."

The recliner creaked. "Oh, now son, you two will work it out. The first couple years of marriage are always a little tricky."

Sam looked up then. "I wasn't talking about the marriage."

Lucy's scalp prickled at Charlie's comment. She addressed Sam. "You mean, moving in here."

"I thought we were doing the right thing. I thought..."

Before he could find the words he needed, Sasha returned. She sat beside Sam on the couch. With one finger under her chin, Sam turned her face toward him. "Do we have some more hard decisions to make?" he signed and said. What he signed must have been slightly off, judging by Sasha's tender smile. She took his hands in hers and formed the correct sign for *decisions*, then spelled it out letter by letter. "Yes?" Sasha asked him with one of the only signs Lucy knew for sure, her eyebrows and forehead turning it from a statement to a question.

Sam had told them *yes* and *no* are key words in their communication. He'd had a crash course in signing, supplemented by Sasha's patience with him, he said, and her excellent lip reading skills. But without the fallbacks of *yes* and *no*, they wouldn't make it through most conversations. "I think you said... Is that what you meant? Yes? No?"

Lucy watched Sam and Sasha work out the miscommunication in a way that made her wonder if the two might teach marriage seminars someday. The newlyweds. The newlyweds whose love looked so different from the expected. So beautifully different.

When Sam's faux pas seemed to have been worked out, Lucy said, "Sam, I don't want to stomp to draw Sasha's attention for fear of waking Evan. Would you let her know I want to talk to her?"

Sam touched Sasha's leg and pointed toward his mother.

Lucy had no hand signals for what she wanted to say. So she enunciated clearly, but at a normal conversation pace, as Sam had taught her. "Sasha, what do *you* want to do? Do you believe we can make this work here as a family until you find a place of your own?"

Sam watched his bride's hands as they flew through a response and did his best to interpret, often having to stop her and ask her to repeat or spell something. Lucy watched Sasha's body language and facial expressions. They told more than the words.

"I've never had a family like this," he said on Sasha's behalf. "I don't believe we have a choice"—she shrugged—"but when God gives us no other option, He also gives exceptional grace."

The thought raked a path through Lucy's mind. She'd lived a "no other option" life since May. Had she been blind or—deaf—to the "exceptional grace" part?

"Okay, then," Sam said. "This is home for a while, and we'll make the best of it."

Sasha signed, "Yes," and nodded her head with finality.

The room quieted. A low rumble told Lucy her husband had found a way to doze off in the middle of the discussion. And another low rumble jolted Sasha. She pulled her phone from her pocket. Vibrate mode. She excused herself and left the room.

"How does she... talk on the phone, Sam?"

"Technology. A godsend for her. For all of us. She has a baby monitor app that allows her to keep an eye on Evan from a child cam. It's set to go off—vibrate—if he gets active even if we can't hear him yet. In Evan's case, that could be a danger."

"A wonderful advancement."

"She loves Skyping with her friends. She doesn't have to type everything, like with e-mail, but can sign or read lips."

Lucy could hear Evan now. He didn't seem happy. "Can I text her?"

"Sure." Sam pulled out his own phone. "She and I do that all the time."

"It would help while I'm still so far behind in learning to sign, and for communicating from upstairs to downstairs. I don't want to interrupt her privacy or flick the lights on and off if I can send a message that asks when she'd like supper."

"Or tell her that the mail's here," Sam said. "Or ask for her help."

"I won't be doing that often. She has enough on her hands."

"How would you handle this live-in situation if Sasha weren't deaf? Do that."

After lunch, Charlie announced he had an errand to run, Olivia met a friend downtown for coffee, and Sam headed to Hudson to put in a few hours at the office. Evan settled into a better pattern with the household down to the three of them. Lucy started a jug of sun tea on the deck. Evan watched for what seemed like hours, fascinated with the swirl of caramel brown that oozed from the bundle of tea bags until the entire glass jug had converted from clear liquid to a little deeper color than what it would be when iced.

Sasha sat with him on the deck, her laptop on the patio table. From the sink in the kitchen, Lucy could see a hint of architectural plans on the screen. What a talented young woman. Life hadn't turned out as planned for her either.

Fighting the too-familiar sense of failure, Lucy did the hard thing and called Bernie and quit her job. How could she juggle that with everything else happening around them? So she'd lost two jobs in the space of less than three months.

"The door's always open," Bernie said. "Except when we're closed."

"Life is a little complicated for us right now."

"No need to explain, Lucy. You do what's right for you."

People really thought that way? Bernie should meet Dr. Hanley sometime.

Lucy grabbed a juice box from the fridge—the natural kind with no artificial colors or flavors, per Evan's needs—and filled two glasses with ice. As soon as she stepped onto the patio, Sasha turned. Lucy wasn't used to living with someone so acutely aware of the surroundings. *No offense, Charlie.*

Song of Silence

She'd covered the juice box with a napkin to allow Sasha to see it first and decide if Evan could have it. Sasha smiled and nodded.

"Evan?"

He turned from watching the sun tea brew, still tapping on the glass with his small fingernails. A recognizable rhythm.

"Would you like...some juice to drink?" She sang to the beat he created and offered the box to him. "Apple juice...is good, I think."

He stopped drumming. She stopped singing. He started again.

"What a lovely...sunny day. Makes a child's heart...want to play." She groaned at her inability to produce clever lyrics on such short notice.

The drumming stopped. He reached for the juice box.

Sasha stomped one time. Evan looked at her, then looked at his feet, not Lucy, when he said, "Please. Thank you."

Lucy had always equated stomping with anger. She had so much to learn.

Evan's voice was sweet, not grating as it was when he was frustrated or out of control. Lucy wasn't yet sure what to call it. Watching him sip his juice and lean his back against his mother while she worked made Lucy's heart swell. Would she ever have a real relationship with this step-grandchild of hers, the kind other grandparents knew? He had yet to look her in the eye or let her hug him.

How had life gone from soul-satisfying to empty to inexpressibly complicated in such a short time?

"Is he hurt?" Lucy stepped around the broken glass but had nothing to offer a child who wouldn't let her near and a mother who couldn't hear her ask. She touched Sasha's shoulder.

Frowning, Sasha waved her off and returned her attention to Evan. The boy had no tears. The blank stare chilled Lucy. He hadn't dropped the sun tea jug in anger, it didn't appear. He'd picked it up and slammed it to the deck with no expression at all. If anything, curiosity. The heavy bottle might not have broken if it hadn't hit the cast iron leg of the patio table.

His mother lifted him from the center of the wreckage and took him down the steps to the lawn, far from the field of broken glass and the tea that had held his attention for hours. She signed to him, held the child who didn't hold her in return, and turned an apologetic face toward Lucy.

Lucy motioned, "I'll get a broom." *Then the Shop-Vac. Then a mop to make sure I got all the shards.*

Most of the tea had seeped through the spaces between the deck boards. What remained made it difficult to sweep cleanly. Each pass caught more pieces of glass within the broom's tufts, which then found a new home on the deck with the next pass. Lucy emptied the dustpan into an empty ice cream pail she'd brought out of the kitchen for that purpose, more conscious than ever of the sound of broken glass. Much like waves on the shore as she swept. More like cheap wind chimes when they clattered into the bucket. She would have talked to herself, as she often did, but Sasha read lips too well. The young woman didn't need to know that every sound Lucy heard made her mourn Sasha's inability to hear it. Even the clattery sounds.

Lucy kept her attention focused on the cleanup, but caught glimpses of mother and son out of the corner of her eye. How do you discipline a four-year-old with Asperger's? She'd had few come through her music program at the school. Some autistic students were provided a full-time aide. The aide took pressure off the teachers to interpret behavior, and deal with inappropriate behavior. Where was Lucy's aide now?

"My help comes from the Lord, the maker of heaven and earth." The psalm, a song, teased the edges of her soul. Psalm 121? She should know this. She'd used that song often in years past when she and Charlie faced tough times. The next line. What was the next line?

"God won't let your foot slip. Your protector won't fall asleep on the job."

She stepped closer to the table to reach the bits hiding in its shadow, the bits that didn't sparkle enough to see without careful inspection. The bottom of her sandal skidded on ambush glass. The shard embedded itself in the sole and acted like traction.

"Won't let your foot slip." Despite the debris of life.

Maybe later, after the mess was cleaned up, she'd find that psalm and read the rest of it. She'd forgotten how the words God spoke into being could snag a person in the middle of an anything but ordinary day.

24

A squeal of old, overtaxed brakes told Lucy the garbage truck had stopped at the house on the corner. Three more stops before their house. Could she beat him to the curb with her ice cream pail of broken glass to add to their week's worth of garbage? She raced around the side of the yard. Quicker than dodging obstacles through the house.

Racing. To an observer, it might have seemed more like an awkward trot. But, breathless, she made it to the chest-high garbage bin on wheels at the end of the driveway just as the garbage truck applied its brakes at the edge of their property.

She greeted the garbage men, thanked them for what they did, and turned toward the backyard to finish repositioning the heavy patio table from where she'd had to move it in clean up. Sasha and Evan stood at the corner of the house. They'd followed her? Of course they had. Sasha wouldn't have heard the garbage truck approach. When Lucy took off at that pace, Sasha must have assumed something was wrong. By the time she felt the truck's rumble, she'd probably raced through innumerable mental scenarios.

"Garbage truck," Lucy said. Big help now.

Sasha nodded that small "yes, I see" or "yes, I understand" or just "yes" response they'd seen often since she and Evan and Sam had moved in. Sasha set Evan down and steered him to the backyard with her hands on his shoulders.

Lucy detoured to the flowerbed in the front yard. The squeal of brakes grew faint as garbage disappeared from the next house and the next and the next. She dead-headed desiccated roses and crispy-brown late-blooming peonies. The tall purple and blush pink cosmos danced undisturbed around her. She'd long ago abandoned making only mental notes of things like aphid spray, opting to supplement her memory with a note on her phone app. Just as she pulled it from her pocket, it pinged. A text message. From Sasha.

"Sorry about your tea jar. Evan doesn't understand consequences."

Lucy thumbed her own message. "Jars can be replaced." Make a mental note to— "I'm glad he wasn't hurt."

Ping. "He's curious but without empathy. He throws things to see what they'll do. Not to break someone's property."

"And he's four." Lucy remembered Sam at four. A mom friend had told her not to be surprised if four looked like a two-year-old squared.

Ping. "Thank you for cleaning it up. Should have been me."

"We're in this together."

A long stretch of text silence made Lucy imagine Sasha embroiled in another scene with Evan, or that the conversation had ended. Do the deaf waste time on the unnecessaries in communication? The endless good-byes? The curious dance of who should have the last word?

Ping. "No one other than Sam has said that to me before. I'm grateful."

He got that from his father. Charlie said it all the time. Today, the phrase felt less claustrophobic and more comforting.

Ping. "Hey, LucyMyLight." Charlie was texting? Texting?

"Yes?"

Ping. "On my way." On his way? Texting while driving. Oh, she hoped not.

"Where are you?"

Ping. "In the mall parking lot. One more stop to make."

Parked. Good.

Ping. "Do me favor? Clear space along south wall family room. Four feet."

"What?"

Ping. "Explain later."

Nothing more.

Her decluttering binge on last year's summer break after reading about the link between serenity and airy décor paid off again now. She moved a lamp table from the left side of Charlie's recliner to the right and relocated a rolling leather-topped two-drawer file cabinet full of music closer to the piano alcove, a better spot for it anyway. Two small adjustments and Charlie would have his four feet.

She resisted imagining what he intended for that spot. He'd given up on anything to do with worms, right?

Charlie pulled in just as Lucy opened the pantry door looking for inspiration for supper. Four extra mouths to feed—counting Olivia—wouldn't necessarily triple their normal cooking habits. Lucy always made enough for leftovers. Evan ate so little, except— Sam said—for items that had a hard time lobbying for inclusion on the food chart. But Lucy now enjoyed a little more freedom in creating interesting salads. Four of the five adults in the house liked salads and a wider range of veggies than the fifth.

Lucy abandoned the pantry scavenger hunt when she heard Charlie grunting. She found him halfway through the front door, an enormous box in his arms, wedged a little too sideways to clear the entrance.

"Charlie, what on earth?"

"Can you get this end? Careful. It's heavy."

Lucy hoisted the end he indicated while Charlie repositioned the other, straightened it out, and then said, "Okay, back up slowly." She did.

"I can't see anything behind me. You'll have to direct me."

"Ooh," he said, eyes twinkling above his end of the box. "Directing the director. I like it."

"Ex-director, and please pay attention. If I fall over backward, whatever this is will land on either you or me."

"A few more feet. Hey, I like what you did with the rearranging."

Now all of a sudden he's noticing the décor? "Losing my grip, Charlie."

"Let's set it down right here." The two coordinated their bend-with-the-knees routine and positioned the box in the space Lucy made.

Lucy had slid her fingers out from underneath just as the full weight thumped its last half-inch to the floor.

"Where are our house guests?"

"Sam's still at work, I suppose. Sasha and Evan are downstairs, I think."

"Good." Charlie arched his back and rolled his shoulders. "I need some time to get this set up. Man, I should have brought the stand in first. That would have been smart."

"I'm getting supper started."

Charlie headed for the front door. "You go ahead. This will take me a while. But I can handle bringing in the rest of the stuff from the car."

The rest of it? Lucy reached to undo the tape on the plain brown box.

"Ah, ah! No peeking. It's a surprise for you, too."

Most of the surprises lately hadn't gone well. Or came with enough baggage to start a used luggage shop. She retreated to the kitchen. Evan sat at the kitchen table with a sketchbook and washable markers. It said so on the box. *Washable.*

Sasha handed Lucy a slip of notebook paper. "Conference call for work. Can you watch him a few minutes? Stomp on the floor if he gets to be too much."

Conference call for Sasha meant typing as fast as she could as if creating her own closed captioning for her boss on the video call. Lucy smiled and made the ASL sign for yes. She smiled in

Evan's direction, too. He didn't notice. She took a spot at the table with him, half for self-protection from messy cleanups and half as a reprieve from thinking about making a meal right now.

"Oh, that's interesting," she said, watching him put random small circles on the page. "Can you tell me about what you're drawing? Can you tell me the story of it, Evan?"

He looked up but didn't quite make eye contact. "Nuffing." He drew circles in the farthest corners, precariously close to the edge of the paper, but...not...quite.

"You must like orange and brown." A lot. He touched no other colors.

No response.

His mother had gotten him a sipper cup of what looked like ice water. He reached for it and took a drink. Then he set it down too close to the edge of the table. Lucy reached to pull it back to a safer spot. Evan grabbed it, as if protecting it from an evil step-grandmother.

"I want to keep it safe, Evan. That's all. Like, here." She tapped on a spot near him but far enough from trouble. "Right here." Tapped again.

He set the cup on the table and inched it closer to the spot where she kept tapping until she had to move her hand. Then he attended to the task of filling the page with orange and brown imperfect circles.

"Nuffing," Evan said, though she hadn't asked a question.

She watched his fingers—midway between baby and child-shaped—clutching a marker in each hand. He drew with his right, but held on tight with his left. The circles stayed small, but she could tell by the width of the lines that he was pressing harder. When the paper held more brown and orange than white, he set aside the markers and picked up his cup.

He tipped the cup over the paper and sprinkled it with water.

"No, Evan!" Lucy grabbed the cup from his hand.

He reached for the cup, whining and kicking his feet. "More!"

Lucy mimicked the calm she wanted him to adopt. "Evan... this is for drinking. Not for messing up your pretty artwork. I'll give it back, if you use it for drinking."

He stopped kicking. Lucy waited. He stared at his paper, but quieted.

"Okay. You can have your water back. It's for drinking. Understand?" She didn't wait for a response not likely to come. "Here." She returned it to the place she'd tapped earlier.

He picked it up slowly. "Tea," he said, flipping it upside down over his paper. "Tea."

The brown and orange watercolor on the paper blended into the color of the tea soaked into the ground under the deck. Was that an apology? Is that what an apology looked like in a four-year-old with Asperger's? Or was the pool of tea-colored artwork flowing toward the table surface an act of rebellion, a statement of his toddler personality?

If Evan hadn't been in her charge at the moment, she would have driven to the bookstore and searched however many sections she had to in order to find the books *When Someone You Love Has Asperger's, When Someone You Love is Deaf, When Someone You Love Moves Back Home, When Someone You Love is Home Too Much, When You've Lost Your Song, Coping with Toddlers, Coping with Premature Retirement...*

Coping. Period.

She pulled a wad of paper towels from the roll by the sink to sop up the colorful river forming on the table. When she turned back from the sink, Evan was gone. So was the paper. She threw the towels toward the puddle and took up the search for Evan. The back door stood open. *Oh, Evan. No! No!*

She found him kneeling in front of the low bench on the deck, patting his artwork into the spot where the glass jar had been. If she'd had her phone, she would have made more than a mental note to put childproof locks on the door. Even if Sam and Sasha found housing somewhere else soon, they'd visit. Lucy began the process of rethinking everything. Everything.

Charlie was already talking about handrails in the bathrooms and wider doorways to accommodate wheelchairs and walkers they'd need someday, as if they were weeks away from being elderly! He could think that way if he wanted to. Early "retirement" was hard enough without living in the land of who's going to get what when we die. Make the house handicap accessible? They needed to make it childproof.

Evan, how are we ever going to protect you?

He hadn't heard her. Her question was internal. But he raised one finger and inspected the dot on its tip. A drop of red. She'd missed a shard of glass.

25

Evan's finger was well washed and bandaged before Sasha came upstairs from the basement. She signed, "What?" and something else, which Lucy took to mean, "What happened?" and pointed at the tea-stained paper towels still in the middle of the table.

Lucy left Evan in the window seat where he'd been since the Band-Aid incident. His attention hadn't left the tip of his finger. But he hadn't complained or resisted Lucy's paramedic efforts. She didn't understand but didn't complain. She grabbed the wet paper towels and soaked up the remaining marker/water, then reached for a fresh paper towel to dry the table.

Lucy stopped mid-swipe to try to communicate that everything was okay now. Sasha signed the now familiar, "Sorry." Lucy took Sasha by the shoulders and mouthed. "No. I'm sorry." She stepped back and signed what she hoped said, "All okay now."

From the expression on Sasha's face and the tender way Sasha placed her hands on the sides of Lucy's face, she judged she might have missed a word or two.

Sasha sat down next to Evan and reached for his hand. He held it out, the bandage on his index finger still fascinating like a candle flame. Sasha asked for paper and something to write with. Lucy recognized the request and grabbed a notepad and pen from the telephone base station.

Sasha wrote, "Evan doesn't feel pain sensations the way most do. It makes it harder to protect him."

Lucy took the paper and pen and wrote, "He didn't react at all to the shard of glass, but seemed taken with watching the drop of blood. And it was only a drop," she added.

"Pain intrigues him. That makes him take risks other children his age might not."

"How scary!" Lucy wondered if Sasha could feel the vibration of Lucy's exclamation point.

"At times, it's good. But more often, it's a danger."

Lucy could sense it in her shoulders and over her heart, a heaviness for this young mom's challenges. She took the pen and paper from Sasha but didn't know what to say. At length she wrote, "How was your conference call?"

Sasha signed, "Okay."

From the family room, Lucy heard a "Hello" at the front door and running water. Running water?

"Dad, what are you up to?" Sam's voice.

She mouthed, "Sam's home" to Sasha and headed toward the sound of running water.

Charlie stood with a garden hose held like a dog's leash, pouring water into—

"It's an aquarium," Charlie said, as if they couldn't tell. "Seventy-five gallons. For the little guy."

Every nearby surface held a filter or light apparatus or bags of aquarium gravel, zippered plastic bags of water plants, roots visible...

"Dad."

"What?"

Sam walked closer to the chaos, stepping over the garden hose that Charlie had strung from the flowerbed through the partially opened front door, through the entry, and across the family room to his project. "This is..."

Sasha with Evan in tow passed Lucy and stood arm-in-arm with Sam. He repeated, "Dad, this is—"

Song of Silence

Lucy caught movement in her peripheral vision. "This is about to get interesting if that chipmunk gets in here!" She rushed the door, no doubt sending the critter's blood pressure way past normal as it escaped back into the freedom of the outdoors.

"Lucy, guard the door!" Charlie said. "I'm going to need another hour to get this tank filled."

Within minutes, Sam had figured out a way to connect the garden hose to the faucet on the powder room sink using waterproof Duct tape and a thick towel collar "just in case." Lucy was off guard duty and back to kitchen duty. Evan didn't say a word, but stood in front of the aquarium, watching his step granddad make like Niagara Falls.

Sasha followed Lucy into the kitchen. "Olivia?" she wrote on the notepaper.

"She has a date tonight," Lucy jotted under the question.

"Caden?"

Lucy appreciated that her kids loved each other enough to converse, but could they fill their mom in on a few of the details once in a while? His name was Caden?

"Help you?"

"Olivia helps when she can. She often works evenings."

Sasha shook her head. "No. Me. May I help you?"

Oh. "Spaghetti and salad?"

Sasha signed something that must have meant "delicious" or "yummy." Her eyes confirmed the intent.

The young woman followed Lucy to the pantry and pulled two cans of San Marzano tomatoes.

Lucy enunciated, "Homemade sauce?"

Sasha signed, "Yes."

Spaghetti night was about to up its game.

Charlie deemed the aquarium "full enough" and his hands "clean enough" just as the women drained the pasta. They found their places around the table. All except Evan, who stood at the door to the deck.

"Evan," Sam said. "It's time for supper. Come sit in your chair."

"Tea," he said.

Lucy and Sasha exchanged glances and reserved smiles.

"That's right, little guy," Charlie said. "If that's what you want to call it. Spa-ghe-TEE."

"Evan." Sam's directive held more oomph this time.

Sasha stomped twice on the floor. Evan turned and made his way to his chair.

Charlie bowed his head, then looked up. "Sam, you want to return thanks this evening?"

"Sure," Sam said and offered his hands to those sitting nearest. Sasha pointed to the Band-Aid on Evan's finger. Sam avoided it as he took Evan's hand in his.

No resistance? Evan allowed that human connection from his stepdad on one side and his mother on the other? Lucy would ask Sam about that later. She took Sam's offered hand and Charlie's. Both felt solid. Familiar. Faithful.

"God of All, we thank You for calling us to this table, where we taste of Your goodness, enjoy one another, and refuel"—he sighed—"with hope for our futures. In the Name of Jesus, our Provider and Protector we pray, amen."

Lucy so wished Sasha could have heard not just the vibrations but the words of that prayer. Did Sasha ever wish the opposite? That hearing people could feel vibrations of prayers?

Evan hadn't bowed his head or closed his eyes, but he hadn't wiggled, either. Not that Lucy heard. She reached for the pasta bowl and passed it in Sam's direction. Sam served Evan first, a small pile of spaghetti. Then he excused his reach over Evan and held the bowl while Sasha dished out her portion.

When the sauce made the rounds, Sam ladled a half-dollar sized pool in an area of Evan's plate far removed from the pasta.

"Ah," Lucy said. "He doesn't want one food touching another?"

"That's right," Sam said. "Not until he...well, you'll see."

Evan picked up a strand of pasta and dragged it through the puddle of marinara sauce to coat the bottom two inches. Then he bit off those two inches and dragged the strand through sauce again. He continued that way until only the spot he held between his fingers was uncoated, uneaten. He dropped the sauceless piece on his plate and picked up another long strand.

"There are easier ways," Lucy said. Something told her Evan rarely chose the easy way for anything.

"Watch Grandpa," Charlie said, twirling his fork into a nest of sauce-covered spaghetti. "See?"

But Evan didn't watch on command. Or by invitation either. They could discipline him with the latest or most ancient methods and not succeed in getting him to look them in the eye. But something would. Something would break through that seemingly impenetrable bubble in which he lived. Something.

Grandpa. Charlie had called himself *Grandpa*. How was it he could move from one reality to another more smoothly than she did? She'd always been the flexible one, and it had nothing to do with her brief stint doing cartwheels as a cheerleader in eighth grade.

Meals were quieter with five or six at the table than they had been with the two of them. Conversation showed its awkward side when the rule of "no talking with your mouth full" morphed to include "no talking with your hands full." Lucy found herself setting her fork aside to chew, in case Sasha asked her a question that could be answered with a *yes* or *no* or one of the few other words she'd mastered in sign language.

If they weren't already married, Lucy would have suggested Sasha and Sam consider it. The tenderness with which they treated each other, the adoring looks they exchanged, the way Sam hesitated a moment when his hand brushed across hers as they passed food platters or attended to Evan's needs. Full-blown love.

"Dessert on the deck?" Lucy asked when all but Evan were finished.

"We have dessert?" Charlie said, incredulity punctuating his words.

"Don't get too excited, mister." Lucy nudged him on her way to the freezer. "Ice cream sandwiches. Store-bought."

"Good enough," he said, pushing away from the table.

Sasha signed that she'd stay with Evan until he was done. Lucy handed two frozen treats to Charlie and Sam, then resumed her place at the table. "I'll stay, too."

"You go," Sasha mouthed and signed.

Lucy shook her head no.

Caring for Evan could be good for her diet.

"Sam told me a little more about Sasha's background," Charlie said as he and Lucy followed their choreographed bedtime routine of teeth brushing and face washing.

"He did?"

"When we were eating dessert on the deck."

Lucy rinsed her toothbrush and tapped it against the edge of the sink. "Glad you two had that time alone. What did Sam say?"

"You'd think a son of mine would be chattier, wouldn't you?"

"Charlie?"

He climbed under the covers on his side. "Sasha had it pretty rough as a kid. She opted to stay in the states when her parents returned to one of those 'zykstans.' I can't remember which one. Can you imagine being on your own in Chicago at eighteen without being able to hear?"

"How did she get through school?" Lucy slathered skin therapy cream on her hands.

"Scholarships for part of it. She's really smart."

"Exceptionally." She layered her pillows for reading and slid under the sheet.

"And...a benefactor where she apprenticed for her architect license. Somewhere near Madison, I think. He asked..."

"What, Charlie?"

"Let's just say he asked more from her than she was willing to give. He took it anyway. Thus, Evan."

Lucy flattened her palm against her heart. The pressure wasn't enough. It broke anyway.

"He abandoned them, of course. And made it rough for her to work as an architect in that area. Especially with Evan's needs. So she moved farther north and wound up at Sam's church."

"And...?"

"You want more details? I thought I did pretty well remembering that much."

"My heart is so heavy for what she's been through."

Charlie punched a valley in his pillow and turned on his side to face her. "My heart feels pretty good about her life from this point on, doesn't yours?"

The vows Sam and Sasha pledged to each other increased in significance, if that were possible. Lucy reached for her book. Charlie put a hand on her arm.

"What? There's more?"

"I thought we should maybe pray together for that little family."

Good idea.

Dr. Hanley shook her head. "Are you sure this is the best thing for you right now? Adding more stressors?"

Lucy sank back into the chair she occupied. She rested her clasped hands on her stomach, partly as a disguise for the two ice cream sandwiches she'd devoured after Sam's little family retreated downstairs the night before and Charlie went back to tinkering with the aquarium parts. "Is there really that much difference between twenty stressors and forty?"

"Every new stressor quadruples the gravitational pull of the others. I'd do the math for you, but—"

"I know. It's a lot to manage."

"How are you sleeping?" Dr. Hanley waited with pen in position.

"Exhaustion and concern fight for dominance at night. Most nights, exhaustion wins."

"I'm guessing you see that as a positive?"

Lucy sat up straighter. "Don't you?"

Dr. Hanley's long fingers pressed the creases out of her forehead. "And your music?"

A familiar pain clawed at her. *Go away. You do not exist.* "What about it?"

Dr. Hanley didn't look up from the notebook in hand.

"What do you mean? My playing? Singing? Listening to it? Are you talking about my finding another career in music?"

Dr. Hanley crossed her arms and said nothing.

"It's a dead issue. Not even on the table right now, okay?"

"Is that what you want to believe, Lucy, or what's true?"

"Remember those forty stressors?"

"Yes." The counselor's voice never wavered from its lyrical, dulcet tones.

Lucy sighed. Her exhale shuddered on the way out. She drew just enough breath to say, "When they closed down the music program, they stole the joy of music."

Was that a look of satisfaction on Dr. Hanley's otherwise beautiful face? "It's taken you a long time to admit that, Lucy."

"I knew that the day it happened."

"But you haven't confessed it until now."

"Confessed? Kind of a strong word. Is that the word you meant to use?"

Dr. Hanley set her notebook and pen aside and leaned toward Lucy. "You just admitted, before God and everybody"—she gestured to the empty room—"that you let someone steal your love of music. It must not have meant as much to you as you claim. Or you would have fought for it."

"No amount of arguing would get my job back." The hair on Lucy's forearms rallied to the discussion.

"I'm not talking about your job. I'm talking about your passion."

"For teaching."

"No." It was Dr. Hanley's turn to sigh. "For the music you taught. The music that you've lived and breathed all these years. The music still echoing in your mind while you slept, the songs that woke you, the lyrics and melodies that rarely quieted down no matter what else happened around you. That. That's what you relinquished without a fight."

He hadn't used the same words, but hadn't Charlie said something like that?

They were both wrong. Both...wrong. Lucy would have told Dr. Hanley as much, but to speak requires air to flow over vocal cords, and the constriction in her throat wouldn't allow it.

"I have homework for you," Dr. Hanley said.

If Lucy cracked a molar grinding her teeth, she'd send the dentist bill to Dr. Hanley or Dr. Hanley's husband, reportedly no closer to clearing the red tape and reuniting their family.

"Two pieces of homework, actually. This. I'd like you to read this. First silently. Then out loud. Then in front of someone you trust. A friend. Charlie. Olivia. Maybe Sasha?"

Lucy took the paper she extended.

"And then, I'd like you to spend ten minutes every day until our next appointment listening to music."

Lucy's groan filled the room. *You don't understand. You can't possibly understand.*

"But not," Dr. Hanley added, "your favorite music. I imagine you haven't been able to stomach your favorite music since May, because of the emotional ache it causes. Am I correct?"

Lucy nodded. She'd give her that.

"New music. New-to-you songs. I don't care if it's baroque or banjo or tribal or reggae or ragtime. Or"—she paused—"worship. Something unfamiliar to you. Take it like a prescription, like medicine, whether it stings or not. Let it work. Ten minutes. That's all I'm asking. For now."

"Music," Lucy said, finally finding her voice, "should never be used as punishment."

26

"There are moments in life, seasons in life, when it seems God goes silent," Lucy read.

She dropped the piece of paper onto the passenger seat of the car. Not now. She'd read it later. When she was...alone.

Okay, so maybe she wouldn't be alone for a while other than this. Here. The parking lot. The hot car struggled to pretend its air conditioning unit didn't mind idling.

She'd give the paper a little more attention. "We wait. We listen. Nothing. No movement. No stirring. No signs of hope or relief," she read. "In the silence, it's tempting to believe nothing is happening." *Tempting? It's a given, isn't it?*

"Good musicians understand the crucial role silence plays in music."

Wait a minute. Who wrote this? She taught it, but who wrote it?

"The pauses create tension. They give weight to what is about to come or what has just passed. These silences are not empty, dead space. They're the crucial, fertile silences that make a song what it is. Musically speaking, in the silence, something is happening."

Lucy flipped the paper to the backside, looking for an author's name. There it was. A guy named Brad. Had he been cyber-spying in her classrooms? Or just particularly astute?

> *In your season of waiting on God to speak, don't be fooled into thinking that just because it seems like nothing's happening, nothing's happening. In the silence, a song is brewing. Even now, in the quiet pause, in the desperate waiting between what was and what is to come, a song is being composed. You are becoming, and this becoming is not empty and passive. It's active, fertile, and alive no matter how hidden it may seem to your ears. Wherever you find yourself waiting...keep composing. Keep becoming. Keep hoping and listening. Somewhere beneath the silence that is your life, a faint chord is striking, a song is brewing.* —Brad Nelson, Restoration Living

Keep composing. Sure. A song is brewing. Right.

"Well, Dr. Hanley"—she folded the paper and stuffed it into her purse—"there's one. I read it silently. And what do you know? I'm not cured."

A song is brewing.

She shoved the thought far from her mind and threw the transmission into reverse.

Ten minutes from home, she risked turning on the radio. She was nothing if not compliant. She could cram in ten minutes of assigned music and be done for the day.

"God, forgive me for my lousy attitude," she sang to the polka tune—new to her—burping from the car speakers.

Her compliance lasted less than a minute. She turned off the radio and drove the rest of the way in silence, except for the sound of molar on molar.

She congratulated herself on graduating to another stage of grief. Anger.

The doorknob felt sun-blasted hot in her hand when she got home. Cool air greeted her as soon as she opened the door. Cool air and more silence. Or... What was that sound? Bubbles?

Charlie and Evan sat with their backs to her in front of a fully operational aquarium. Charlie perched on an overturned five-gallon bucket. Evan sat on the ottoman beside him. No one spoke, but Charlie raised his arm slowly and waved his fingers backwards to welcome her, then returned all his attention to staring at fish.

He'd bought tropical fish. How much was *this* new passion costing them?

Evan's thin shoulders rose and fell in an unhurried, serene rhythm. Lucy was not about to disturb that scene. She counted a dozen fish from where she stood. A colorful mix. Even treading water, they were beautiful.

Even treading water.

Lucy started to turn toward the kitchen but noticed movement. Evan put his hand on Charlie's back. Small hand. Big back. Huge moment. Lucy couldn't see Charlie's face. She could only imagine its expression. But she could see him bow his head.

Her tears were perfectly legitimate. No need to rein them in.

Or Sasha's either. She stood in the doorway to the kitchen, leaning against the frame, watching the scene with her fist pressed to her mouth, face scrunched, eyes glistening. Lucy joined her and put her hand on Evan's mama's back.

A new sound? Behind them. A pan boiling dry.

Lucy grabbed an oven mitt and pulled the pan off the heat. The contents of the pan bore a faint resemblance to beans. The first picking from the second crop of Kentucky Blue Lake from the garden? Lucy slapped a lid on the pan to keep the scorched beans from smoking up the kitchen.

Sasha was at her elbow, talking furiously with her hands. Lucy missed most of the words, but recognized *Sorry*. Over and over.

With the stove turned off and the immediate danger thwarted, Lucy faced Sasha. "It's okay. It's all okay."

Sasha shook her head and signed *No!* Brow creased and mouth set firm, she reached for the ever-present notepad and pen. She shook the pen and tried again. Dry. She slapped her hands to her sides and dug in her pocket for her phone. Sasha thumbed a message then handed it to Lucy.

Lucy read, "I know better. Others cook with sound. I cannot walk away from a pan. So sorry."

Cook with sound? Lucy did, didn't she? She heard the toaster pop up. Heard the unique pitch of boiling water. Heard microwave popcorn slowing. Heard water running, oil spitting, the sizzle that showed the grill was hot enough to make a good sear.

She walked away from the stove often to load or unload the dishwasher or start another element of the meal, relying on her sense of hearing to warn her when the sizzle reached a fevered pitch or the potato water was burbling over the top of the pan. Sound alerted her to so many potential dangers. How different cooking must be for Sasha.

Lucy tapped a message. "Sasha, no worries. We'll find ways to work around this."

Sasha took the phone from her, forced a smile, then thumbed, "I know better. I cannot walk away."

Lucy replied, "Would you have wanted to miss the scene in the family room?"

"No."

Lucy wrapped her mitted fingers around the saucepan's handle and carried it to the back door. Releasing the smoky contents into the outdoors would have fewer lingering consequences. No one in the house had lost their sense of smell.

When she came back in, Sasha met her with a message already keyed into her phone. "Could have been so much worse! What if you hadn't come home then?"

Lucy tapped out, "God knew where I was needed. And when."

He did? He did. Exactly when. Her heart clenched. He knew she'd read the words about the song He was creating in her silence only one time then cram it into her purse. He knew.

And used her anyway.

"It is good to wait in silence for the Lord's deliverance."

No fair, God. I finally open my Bible and this is what You slam me with?

Lucy let her eyes scan farther down the crisp page. Lamentations. Chapter 3. *What else do you have?*

"He should sit alone and be silent when God lays it on him."

She closed the book. Then opened it again. But her thoughts strayed to how few words she and Sasha exchanged as their relationship developed. A look. An expression. One mouthed or signed word communicated so much. Enough, with a few important exceptions.

She'd thought her connection with God depended on the words she said to Him or for Him to others, or the volume of words she read about Him. Maybe she'd been approaching this too noisily. Maybe what her heart craved was more about the silence, the rests, than she'd imagined. Few words. *Let my words be few.* Even that came from somewhere in the book she held, didn't it?

Lucy sat late into the night on the window seat in the silent kitchen, the only illumination from the moon and the small bookshelf lamp above her head. A swirl of thoughts played like beginner saxophone notes in her mind. *This will never work. None of it.*

It has to work. All of it.

As a teacher, she'd built a reputation for confidence and innovation, for discovering teaching techniques that reached the unreachable and offered struggling students an outlet for their emotions and creativity, and a reason to celebrate. She tapped into the math of music, the science of music, its drama, its physiological components, its mechanics, its cultural and social connections.

And they responded. The students responded. To the music. To her. To the idea that music opened them to a world of enjoyment.

But here, in this new old environment—her own home—nothing was as it should be. Nothing promoted confidence. Charlie seemed a little lost. She was a lot lost. Sam hated to go to work but had to. Sasha—so bright, so skilled, brilliantly hurdling over obstacles the world imposes on the deaf but mired in the maze of her son's disconnects. Did they feel abandoned to the basement dungeon-on-its-way-to-living-quarters, as Lucy would have? And Olivia—stuck with her parents in a terrarium of psychological oddities about which she was studying. A laboratory.

Their family certainly offered Olivia plenty of material for case studies.

The door between the kitchen and garage creaked open.

"Mom. I didn't expect you to wait up."

Lucy swung her legs over the side of the window seat. "I wasn't waiting up, exactly." Just waiting. She stifled a yawn. What time was it? After midnight? "Late night at work, huh?"

"Mmm, that was part of it." Olivia joined her on the window seat and grabbed a throw pillow to hug.

"What's the other part?" One of these days, her daughter would tell her to mind her own business. Lucy couldn't afford the emotional energy to mourn the coming of that day.

"So you know I've been dating this guy."

"I believe you may have mentioned it. Caden?"

"I don't remember telling you his name."

Lucy cupped her chin with her hand. "No. I had to find out from Sasha."

"So you know then."

Where was this headed? "His first name. Is that all I'm allowed to know?" Her open-book daughter had something to hide? No. God in heaven, please tell me I'm imagining things. It's late. That's it. Overtired. Not on my game. Inventing problems that don't exist.

Lucy hadn't played guitar for months. But she remembered the snap and sting of a broken string. Somewhere in her core, the place that registers tension, a string broke. She was pretty sure she didn't have a spare in the case or the time to replace it. And the sad thing was Olivia's words should have made her heart sing. Lucy would have to play with a string missing. "Tell me about him. Want me to make some decaf?"

"Coffeed out."

"Talked out, too?"

Olivia's smile would have made her face glow even without the summer moonlight and the small lamp. "No."

"Oh, my dear child. Are you smitten?"

She shrugged her shoulders and clutched the pillow tighter. "Perhaps a wee bit."

"Was he the one you left with after the budget meeting?"

A completely open-hearted answer comes within a second of a question like that. Lucy counted four seconds. "Yes."

"Reporter?"

"No." Olivia shifted to a more upright position. "He's running a program at the Performing Arts Center. Amazing guitarist. Caden Cole."

That didn't explain why he attended the community budget cut meeting. "Olivia! I need to know more than that!"

"Mom, shhh. You'll wake the troops. I didn't say we were engaged or anything. We...talk. A lot. About everything. As I said, he's a struggling musician."

Aren't we all? Different reasons. Interesting. Olivia started her description with what this young man did for a living. The HHATT book club discussed that the week before. When a man—or woman—retires, what happens during introductions? Are they identified by who they used to be?

She'd introduced Charlie with, "My husband, Charlie. He's retired." Or, "He's a retired mill worker."

The HHATTs discussed a different approach. Lucy had yet to get the opportunity to try it. Introducing someone by their best character qualities. Who they are, not what they did or once did. "This is Charlie, my husband. He's devoted to his kids and his grandson, and to me. He has a generous heart."

What kind of conversations might that engender at a dinner party?

"Mom? You still awake?"

"Thinking. About this Prince Charming of yours. All I know so far is what he does. What is it about him that made you want to know more than that?"

"Ooh. Interesting question." Olivia massaged her scalp and fluffed her hair.

"The dreamiest eyes you've ever seen?"

"So cliché." Olivia brushed off the thought with two hands then closed her eyes and draped her wrists on top of her head. "But they are."

"Hate to tell you this, but that's not enough to build a relationship on."

"I completely agree. It's everything else about him..."

"Should I be taking notes for you?" Lucy scribbled in the air.

"That's okay. I have him memorized."

Gone. Er. Her daughter was a goner. "Can you tell how he feels about you?"

"Well, we have another date Saturday."

"Great."

"We're"—Olivia feigned nonchalant—"looking for rings."

Lucy sucked in air so fast she choked on it. Coughed. Kept her mouth closed while her windpipe settled down.

"Onion rings, Mom." Olivia's sly smirk gave away how much she enjoyed Lucy's momentary discomfort. "He's as much a fan of them as I am. He's toured the Midwest..."

Oh. That kind of musician.

"...and stopped along the way at every place famous for its rings. There are two restaurants not far from here, so we're going together."

"Sounds like an odiferous date. About as much fun as a garlic tasting."

"He's done that, too. But not here. When he toured out in Gilroy, California. Did you know the freeway smells like garlic there certain times of the year?"

Lucy adjusted her own pillow. "I can see you two have had deep, deep discussions already."

"Isn't it a sign or something if you can talk with a person for four hours nonstop and never get bored?"

Since the family moved back home, Charlie and Lucy had maybe three conversations that lasted more than a couple of minutes. Isn't that a sign? "I know you're excited, Olivia. Just—"

"I know what you're going to say. 'Take it slow, little missy.'"

"I wouldn't have used 'little missy.'"

"And I know that's good counsel." Olivia put the pillow behind her back and hugged her knees to her chest. "But I haven't felt this way . . . ever."

"Fee-lings . . ." Lucy sang intentionally off-pitch.

"Come on, Mom. I'm being all serious and vulnerable here."

Lucy pinched her finger and thumb together and used them to zip her lips. She unzipped them to ask, "Caden Cole. Is he a musician I should know?"

"Not famous. Yet. He's working on building additional programs for the Barre Performing Arts Center. I told you I could have gotten you a ticket for tonight."

"The classical guitarist?"

"Yes."

Ironic. One of these days, Lucy would have to find where she'd misplaced her courage under all that humiliation and go to a concert or two while Olivia still worked where she could get discounts.

"When will I get to meet him?"

Olivia's quirky grin showed up again. "Let's see how we do shopping for rings first, okay?"

Sometimes Sam and Olivia were a little too much alike.

"I want you to meet him. It's just..."

"Olivia, you're torturing me."

"He's afraid to meet you."

What kind of young man is afraid to meet the parents? Not good. Not a good sign. "We're not that scary, honey." Lucy ran her fingers through hair with a visible aversion to humidity. "Not *that* scary."

"Mom." Olivia stared at the pillow in her arms before lifting her head. "He's Evelyn Schindler's son. Cole is his stage name."

Lucy heard the sound of an entire shop of guitar strings snapping.

"Now you know why I've been hesitant to tell you." Olivia tapped her fists together as if playing rock-paper-scissors with herself.

"I don't know what to say."

"I didn't think it would matter, unless Caden and I... kept seeing each another. Mom, he's nothing like Mrs. Schindler."

"Still don't know what to say."

"A half-truth is as bad as a lie. I'm sorry for that, for not telling you everything right away. I wasn't sure how you'd take the news, Mom."

The heartburn morphed into heartache. Olivia couldn't trust her mother to act like an adult, a well-adjusted, emotionally healthy, non-grudge-bearing grownup. "You thought I was too fragile?"

"I wanted to be sensitive because of what you've been through. Now I know that was disrespectful."

Lucy rubbed the muscles at the base of her neck.

"Mom?"

"Come here, honey."

Olivia leaned into Lucy's embrace.

"If you're impressed by Caden Cole, I will be, too. I'm sure of it."

What she couldn't understand was how Evelyn Schindler could have given birth to a musician.

27

Clang-clang-clang-clang-clang. The sound stopped, then started again a millisecond later. Clang-clang-CLANG.

Lucy opened her eyes and oriented herself. Bedroom. Sun shining. She unburied herself from the plethora of pillows and snuck a look at the alarm clock. A little after seven. A reasonable time to get up if she and Olivia hadn't talked so late.

Clang-clang-CRASH!

Her piano. Lucy threw off the covers and wrapped herself in the throw that hugged the foot of the bed. Charlie must have showered already. The humidity in the bathroom had reached Amazonian levels. Discovering the source of the piano noise—she had a prime suspect—had to wait until she had a moment in the steamy bathroom.

She exchanged the throw for a terry cloth hoodie she kept on a hook behind the door and bolted for the family room.

The noise had stopped. Evan sat on the piano stool, staring at the keyboard lid that rested on his fingers. That was the crash? The piano lid had slammed on his fingers?

"Evan, honey, let me take a look." She lifted the lid and snugged it back into its open position. Six small fingers were dented. He ignored what seemed scream worthy to Lucy and started pounding on the keys again.

She didn't remove him from the piano, but lifted him enough to slide him onto her lap. "Let's try this," she said. She cupped his right hand with hers and folded down all but his index finger. Hers rested lightly on his as she started at the beginning—"Twinkle, Twinkle Little Star." Or "The Alphabet Song," depending on one's mood.

He'd stiffened when she took his hand, but relaxed by degrees as she guided it to make music. Simple music, but a pattern he must have recognized. "I do again," he said. His longest sentence since Lucy met him.

He didn't kick against her presence. He waited for her hand to guide his over the keys. By the sixth or seventh round, she felt his fingers aiming for the right key before she moved. Not consistently, but enough to give her hope.

Evan resisted when she switched to "Mary Had a Little Lamb." He muscled back to "Twinkle, Twinkle." Under the circumstances, Lucy deemed it prudent to give in and let him take joy in what he knew.

He shrugged off her guidance and attempted the song on his own, managing the first phrase over and over. And over. After several minutes, she reached around him with two hands on the keys and began her favorite version of "Jesus Loves Me." Evan no longer leaned over the keys but sat upright, leaning against her. As she played, he laid his hands—light as down—on hers. She slowed the tempo so he could feel what she felt—the magic of coordinated finger movements producing song. Dynamics and nuance producing musicality.

Lucy changed keys, changed styles. Evan stayed with her. This time, she didn't try to change songs. As much for her as for him.

She kept the music simple, for fear those small warm hands would slip off hers and not return. Same song. Evan's bare heels tapped the rhythm lightly against the front of her calves, pausing on the rests.

Bliss. A sweet moment of bliss. This is what she was born to do—connect to someone's soul through music. She'd thought that meant a sea of children, a new wave every school year. But here she sat—not even ready for the day—coffeeless, shoeless, connecting with one child. One broken child.

Lost in the wonder of Evan on her lap, caught in the current of an ancient melody and timeless truth, she began singing the lyrics. *Oh, Evan. I hope you don't mind morning breath.*

In the background, she heard footsteps in the kitchen. Olivia? Sasha? It was Sasha. She approached the piano from the side, where Lucy could see her. It must have been ingrained in her to do all she could not to startle the hearing impaired or deaf. Lucy needed to learn that lesson. She hummed rather than sang now, a fragment of her attention diverted to the expression on Sasha's face. Had she not known where Evan was?

Sasha looked newly awakened too. Maybe Evan had found his way upstairs while she still slept. She seemed captivated by Evan's fascination with the music, with the calm in her son. She watched, smiling.

Eventually, Sasha touched the piano. First with her fingers, then with her palm flattened on the baby grand's closed lid. She laid the side of her face against the surface, not hearing but feeling. She righted herself then and rubbed with her sleeve on the spot where her cheek had been.

When Sasha looked her in the eye, Lucy mouthed, "Lift the lid. Go on."

The young woman opened the lid, found the topstick, and propped the lid open at its highest point. She laid her hands so lightly on or above the strings that the sound wasn't noticeably muffled. Lucy kept playing.

Sasha's face brightened. She signed, "Jesus Loves Me?"

How could she tell now that Lucy had stopped singing? She couldn't be reading lips. Was she that familiar with string vibrations or piano keys that she could recognize a pattern and connect it to a song?

Lucy almost wished she could go back to school and study the connection between music and the deaf. But that sounded too much like dreaming. She wasn't ready for that.

But there came that familiar chorus again.

Evan's serenity didn't last long after the music stopped. Lucy tried holding his hand for the notes of "Twinkle, Twinkle" again. But he scooted off her lap and dove headfirst into the couch.

"I guess I'll go brush my teeth," she said, gesturing with an imaginary toothbrush.

Sasha signed, "Me, too." She pulled Evan from the couch and tugged him toward their basement hideaway. At the doorway, she paused and turned. "Thank you."

Lucy didn't know how to sign, "My joy." Before the day was over, she intended to learn.

She left the piano lid open as she headed for the master bedroom. No real reason. Within minutes she emerged, more ready to start her day than she had been in a long time. Charlie met her in the family room.

"I closed the piano lid for you," he said.

"Oh. Okay."

"Don't know who opened it, but I'm pretty sure you don't want Lego bricks and Goldfish crackers flying in there accidentally. That Evan's got great aim, when it's something heavier than a paper towel."

The wedding incident.

"So, what are we going to do today?" Charlie clasped his hands together as if he'd been waiting for the opportunity to ask. "Now that, you know, you're *really* not working."

Lucy had homework for both Dr. Hanley and the HHATT club. Charlie knew about Dr. Hanley, generally speaking, and had only once asked, "So, how many sessions do you think this is going to take?" He knew she attended a book club, but had no idea how significant it was in her ability to fend off resentment for things like his inadvertently snuffing the emotional echoes of the morning's music when he closed the piano lid.

Homework. And the relentless housework. But the outside temps were on the mild side and extended time alone with Charlie was long overdue. Besides, he probably hadn't heard the Olivia Meets the Guitarist of Her Dreams story. "Did you have an idea?"

"Oh, I don't know. Did you?"

"We haven't canoed the Willow at all this summer."

"Are you up to that?" Charlie asked.

How would emotional upheaval and losing her music program and having a house bulging with noise, lack of it, and enough baggage to start an airline make her unfit for canoeing? "I'm still younger than you, Charlie."

"Prove it," he said, and took off for the kitchen at a pace he called hurrying.

She followed, amused at the reminder that she'd once asked if he couldn't hurry. He'd said, "I only have two speeds. If you don't like this one, you're sure not going to like the other one."

The ability to make his wife laugh was a good quality in a man. She'd remind Olivia the next time she saw her.

Charlie was quickly earning a reputation as the Sandwich King in their house. He layered sliced ham, pickles sliced the long way, and provolone on rye bread he'd slathered with something he called a secret ingredient. It looked a lot like mayo and mustard stirred together. Lucy pulled a soft-sided cooler from the pantry and filled it with fruit and veggies and two bottles of water.

"Sunscreen?" Lucy asked.

"In my tackle box."

"Oh, we're fishing, too?"

"Might as well, while we're there."

She couldn't fault him for not being able to read minds. The HHATT club girls reminded each other of that truth frequently. "Charlie?"

"Yeah?"

"Do you mind if we don't fish this time?"

"Why wouldn't we?"

Like a teenage girl picking an outfit for the school dance, Lucy tried on several responses and tossed them all aside. What would he understand without feeling hurt or disappointed?

"So, you don't like fishing all of a sudden?" Charlie said.

"No. I like to fish. I hoped you and I could..."

"Could what?"

"Talk. Float on the river."

"We can do that for a while. Sooner or later, we have to paddle."

Lucy wrapped her arms around him from behind. "I know. When we're fishing, though, that's where our attention is. And we have to be quiet."

"You were planning on being noisy?"

"Forget I said anything."

"No. No. If you don't like fishing anymore..."

Lucy released her grip. "Charlie, that's not what I said."

"Maybe you'd rather not go at all."

She stomped her foot. "It was my idea."

"So it's all okay, as long as it's your idea."

"Charlie, wait a minute. How did this derail? We were planning a nice day together on the river."

Sasha opened the basement door. "What is it?" she signed.

She couldn't have heard their discussion. Lucy and Charlie exchanged glances.

Sasha stomped one time and pointed to her feet.

"Oh." Lucy stepped closer to her. "Sorry," she signed, reenacted a bit of the scene, then put her hands up in surrender, which probably meant something completely different than, "Mea culpa. We were being childish."

Sasha must have understood enough. She grinned while pulling the basement door shut again.

"I need a hat," Lucy said. And the HHATT club.

"I do, too," Charlie acknowledged. "Meet you by the garage? I'll get the canoe on the car rack."

"Great." She kissed Charlie on the cheek. "This will be fun." Fishing or no fishing.

Lucy fielded two phone calls and let a third go to voice mail. She finished loading the dishwasher and let it start doing its thing. She made a quick trip to the bathroom—no facilities out there on the river. By the time she'd found the waterproof shoes she wanted to wear, she expected Charlie would have gotten the Traverse loaded with their fishing gear, life vests, net, the tackle box, and secured the canoe on the rack. And that he'd be tapping his foot while he waited.

She had other people—a whole houseful—to think about, now that she and Charlie no longer lived alone. The basement progress and apartment hunting was slower than they'd dreamed it would be because of Sam's long hours and long commute. It would all work out. Somehow.

Lucy dug some frozen meatloaf from the freezer to thaw for supper and jotted two more items on the grocery list.

She might need reading material if they floated, or if Charlie found their lunch stop a good time to sneak in a little nap. She grabbed a couple of magazines from the rack in the family room. One had been bent open to the next to the last page. How long had it been since she read a magazine? This one might have been sitting in the rack since Christmas break last year.

The opened page caught her eye, though. A serene scene. A river lined with willows. She'd have to show that to Charlie. The photograph was overlaid with text that said, "Open your mouth only if what you are going to say is more beautiful than silence. —Arabic proverb."

Her lunch would consist of a Charlie specialty sandwich and food for thought.

28

Lucy left a note for Sasha, then decided to text her, too. To be on the safe side. Communicating without sound proved challenging, but so much easier with the aid of technology. Every day, Lucy noted a new benefit, too, to letting actions do the talking. When Sasha spoke, Lucy didn't merely hear words. She watched the woman emote.

Like a violinist using body language as well as his bow and finger positions, or a singer exhausted not from the effort, but from the depth of expression. Who would have thought Sasha's presence in their lives would stir Lucy's thoughts the way they had, would reshape her perceptions?

Olivia's door remained closed, no doubt still dreaming about the guitarist she found so captivating. What mother would want to disturb a dream like that just to say, "Your dad and I are adventuring today. Canoeing the Willow. See you later"? So Lucy texted her, too. Then she dug through the shelf in the hall closet for her lime green HHATT club visor and lightweight shirt-jacket she didn't mind getting dirty if they did wind up fishing.

Charlie was cinching the last of the tie-downs for the canoe as Lucy came through the garage with the lunch cooler. She saw life jackets but no fishing equipment in the back end of the car. Should she say something or let it go? Apologize? Stand her ground? Restart the discussion?

She let it go.

"Let's both drive," he said. "Then we can leave your car near the boat ramp at the park in town and drive the Traverse up past Calder's Ridge. That'll make a nice long float. No paddling against the current. When we land at the park, we'll drive back up to the Calder's Ridge put-in point to get this vehicle."

"My car doesn't have a canoe rack."

"Oh, yeah. Well, you can wait there with the canoe while I drive to get my car with the rack and leave your Malibu at Calder's. Then when we get the canoe loaded on the Traverse, we'll drive together to Calder's to get the Malibu and then—"

"This is getting so much more complicated than I hoped."

Charlie removed his cap and wiped his brow with the back of his hand. "So, do you want to go or not?"

The chaos of their compounded crises was taking a toll on him, too. She saw it in his eyes. The tension in his voice confirmed it. He'd wanted her and her alone for this next stage of their lives. Not only was his wife not content with retirement, or the rude way in which it happened, but their aloneness now included a daughter, son, daughter-in-law, and step-grandson. And a gargantuan aquarium.

"Yes," Lucy said. "Let's just go and figure it out as it happens."

"Who are you and what have you done with my real wife?" He gave her a quick hug and settled himself behind the driver's seat. When he'd backed his Traverse onto the street, he pulled to the curb until she'd gotten her car out of the driveway, too. He usually chose to follow her if they drove somewhere separately. Protective, she supposed. He powered down his window and waved her to go ahead. Like she needed his protection to get from their house to the park downtown.

They left her Malibu in a safe spot in the parking lot. She locked it and put the keys in the cooler. The key fob could fall out of a pocket too easily. Lucy climbed into the passenger seat of Charlie's car and noted that his radio wasn't on. No political or news commentary would accompany their trip. So much she'd taken for granted in life had been background noise.

"Well," Charlie said when she'd buckled in, "maybe we do need to keep both cars. So we can do stuff like this."

A couple of weeks ago, he'd hinted that they'd save a lot of money if they revisited the idea of becoming a one-car family. Lucy's hair straightened itself at his comment. He'd changed his mind again? She slipped her visor on in preparation for whatever came next. For recreation, he'd be willing to keep two vehicles? But not for convenience. Or her freedom to come and go without working around his schedule?

But that's what husbands and wives did, didn't they? Adjust? Work around each other? Or was every marital ride supposed to be on a tandem bike? He turns left, she turns left. He brakes, she brakes.

At the moment, in their version of "Nineteen Kids and Counting," there might be as many as four vehicles in their driveway. Olivia's, hers, Charlie's, Sam's... Sam and Sasha perused the rentals and real estate ads almost every evening looking for an apartment or condo or small home. Nothing suitable so far. Evan's unique concerns coupled with Sasha's unique concerns coupled with Sam's commute made the search as complicated as trying to get two people, a canoe, and two cars where they were supposed to be.

The traffic jam in Lucy and Charlie's driveway could ease by a quarter with the announcement that Sam and Sasha had found a place to live. The thought emptied her for some reason.

How involved would she and Charlie be once the family moved out? Were they the "adequate daycare" Evan needed? How would Sasha handle getting Evan to and from school? Sam said many deaf people drive, but the young woman seemed hesitant to work toward getting her license. If Charlie kept connecting with Evan the way he did, almost as if he understood how Evan's brain operated, and if Lucy could connect through music, is it possible they could be more than step-grandparents? They could form part of Sasha and Sam's answer.

"You don't like that idea, I guess." Charlie's voice held an air of defeat.

"I'm sorry. Lost in thought. What idea?"

"Keeping both cars."

Hadn't that been her position all along? Sometimes she wondered... "I'm all for it."

"As long as we can afford it."

"Right."

"Our taxes are going to be a mess to figure out this year."

"Charlie?"

"Yes?"

"Let's talk about something else, okay?"

He glanced at her, then gave the road his full attention again. "Pick a subject. Any subject."

"Are you ready for this one?"

"I don't know. Try me." He glanced down at his feet, for some reason. "Wait. You're not pregnant, are you?"

"Very funny. No. Our Olivia may be on the cusp of a serious relationship."

"Get out!" Charlie pulled the car to the shoulder of the side road and checked his side mirror while he jammed the shift stick into park. "Get out!" He fled—was there another word for that kind of speed?—the vehicle.

"What?"

"Lucy!"

He wasn't kidding? She opened her door and slid out, teetering on the uneven shoulder that sloped to a weed-filled ditch. She watched through the open door as Charlie bent down and gripped the driver's side floor mat protector with two fingers of each hand. He pulled it out of the car and let it slam onto the concrete while he danced his way to the back of the vehicle. She joined him there.

"Charlie, what's going on?"

"Garter snakes do not,"—he leaned one hand on the canoe tie-down that led to the trailer hitch—"belong in automobiles. Period."

"Oh, you've got to be kidding!" Lucy snuck a peek around the corner of the car. A foot-long garter snake the diameter of a pencil lay squiggled on the floor mat. "It's not moving."

"Well, it had better or we're leaving the floor mat right where it sits."

"What if it's a baby?"

Charlie bent his neck to look. "Probably is."

"What if its mom is still in there?" Lucy couldn't keep the amusement from her voice. The mighty almost-worm-farmer panicked at the sight of something not much bigger than a healthy, well-fed earthworm.

"That's it. We're selling the car. As is." Charlie scuffed gravel as he turned and marched off.

"You're walking back to town?"

"It'll be good for me. Keys are in the ignition. Meet you back home," he called over his shoulder.

She stared at his retreating Ed Asner back. It bent forward. And stayed bent. Lucy rushed toward him. She peered both directions down the side road. No cars. "Charlie!"

How could she have been prepared for what she saw when she reached him? His body convulsed. With laughter.

"Charlie William Tuttle!"

"Okay," he spit out between guffaws, "in my defense, I didn't plan this. And that's a real snake and it really did get my heart racing."

"You are too old for practical jokes." Lucy's blood pressure kept rising. She could feel it.

"I told you, it's a real snake. I don't know how it got in there. But hearing you tear after me like you thought I was dying...? Priceless."

When given the choice, always opt for humor. Week five of the HHATT club journal entries. *Find the humor in the mishaps, the mistakes, and the mayhems. Your blood pressure will be grateful.* That had come from Angeline, if Lucy remembered correctly. Angeline had read it in a book she'd picked up at the library. The idea had sparked an hour's worth of discussion.

Lucy pulled out her cell phone.

"What are you doing?" Charlie asked, hands on his knees, breathing hard.

"Recording this moment for posterity. And Facebook." She snapped a picture of the car with the canoe on top and both front doors open, the floor mat on the asphalt. She caught an image of her husband bent over, sucking air. And she returned to get a close-up of the garter snake, still squiggled in the same position. Huh. "Charlie, come here. Does this thing look dead to you?"

Charlie drew close enough, barely, and said, "All snakes are 'dead to me.'" He flicked his hands to the side and turned his nose to the air.

"Well, my big, strong hero, this one has been dead for a while. Either that, or you frightened the liver out of it when you threw its bed to the asphalt." She kicked at its skinny tail with the toe of her shoe. It slithered off across the road to the opposite ditch. Both adults jumped back. "Okay, so it was merely meditating. Now we know."

The two erupted in laughter again, clinging to each other like longtime friends. Friends.

When the humor of it all faded, Charlie said, "We're going to have to get back in that car, aren't we?"

"Afraid so."

"You first."

"It's obvious this little guy was a loner."

"What makes you so sure?"

Lucy made her way to the passenger side of the car and climbed in. "Did you see that suit he was wearing? Mismatched colors. Definitely no woman in his life."

"Nice one."

"Thanks." Of all things, the song that came to mind was a Smothers Brothers' tune from decades ago, one her seventh grade boys found hysterical when she played an LP in their "history of sound" unit. Yes, definitely time to break into a verse of "The Slithery Dee."

Charlie found it less entertaining than she did. But he looked under the seats and, satisfied for the moment, slid behind the steering wheel. They were a mile down the road before he said, "What's his name?"

"I thought we'd remember him as Dee. Or, Slithery, if you prefer." She thumbed her phone and sent the images to both kids.

"Not the snake. Olivia's heartthrob."

"We didn't get to that. Oh, yes we did. Caden." She affected a starry-eyed pose. "Dreamy name, isn't it?"

"As long as he treats her right, I don't care if his name is—" He cleared his throat. "Yeah, can't think of anything funny right now."

"I've got something," she said, and pointed the phone screen image of the snake scene toward him.

"You didn't really send that, did you?"

"Oh, but I did."

29

It was close to noon when they got to the water. And close to ninety, according to the temp on the car dash. Lucy entertained a brief thought that a full hat might have been a better choice than a visor. The cooler air of the early morning—and the weathercaster's predictions—had both abdicated their reigns.

She squinted against the high sun as she helped slide the canoe from rack to ground. "We'd better lay the sunscreen on thick, Charlie."

"About that..."

"Char...lie?"

He handed Lucy her lifejacket.

"Where's the sunscreen?" Her singsong tone held a barely restrained irritation.

"In the tackle box," he mimicked back to her.

"Which is...?"

"Back at the house. We decided—*you* decided—we weren't fishing today."

Resentment fouls relationships like rotted potatoes. Marta's HHATT journal contribution. "I love you soooooo much," she said, "that I'm giving you a pass for that tone of voice."

"And I love you sooooo much," he parroted, "that I'm giving you a pass for yours."

And this was what an enduring marriage looked like. Giving undeserved passes for the relatively small things as practice for giving undeserved passes for the big things.

Practice. Piano scales. Vocal warm-ups. *May-mee-mah-mo-moo.* Triplets—*pineapple, pineapple, pineapple, pineapple.* Arpeggio exercises. Triads, higher by half steps. Lower by half steps. Performance posture even during rehearsals. Never a time when excellence is not the goal.

Practice for the harder things. Building muscle memory so the fingers or vocal cords or embouchures know what to do without thinking...which enables musicians to advance beyond capabilities.

Undeserved passes. Practice.

Charlie lowered himself into the rear of the canoe while she steadied it. "Careful getting in, LucyMyLight. Slithery rocks."

She quirked an eyebrow.

"*Slippery* rocks." He held the canoe close to the shoreline with his canoe paddle dug into the bottom of the shallows on the far side of the vessel.

She'd been entering canoes since before they were married. He'd been telling her to be careful at least that long.

Hot as it was, when they pushed away from the tree-shaded shoreline into the heart of the Willow River, they felt a fuller impact of the sun's enthusiasm. They could stick to the spotty shade along the shore and miss the benefit of the current or tough it out in the middle. "How far are we from town, Charlie?"

"By river? You know the Willow. Crooked as an old man's small intestine."

"Charlie!"

"By road, it was seven miles."

"Okay." Lucy pulled her life vest away from her neck. If she'd been a stronger swimmer, she would have sat on it, rather than wearing it, which would have solved two problems. She'd be cooler and the canoe seat would be decidedly more comfortable. The Willow wasn't all that deep, but deep enough.

"By river? We're probably talking twelve."

"Twelve miles?" She plied her paddle as he'd taught her long ago, but not digging in far. The current did most of the work.

"About that many."

"And how many miles per hour?" She wiped a drizzle of perspiration from her hairline.

"The current's probably running, oh, two."

Lucy stopped paddling and swung the upper half of her body to face him. "Charlie, that's five hours on the water."

"It's true what they say. Musicians have heightened math skills." He chuckled.

"Charlie, we cannot be out on the water for five hours in this heat and sun, without sunscreen."

"That again?"

She sighed. "I didn't mean it unkindly." Not consciously unkindly, anyway.

Charlie motioned for her to turn around. "We don't have much choice. Do we?"

"What if we paddle hard?"

"We could up our speed to... maybe... three miles an hour."

No matter how she did the math, she couldn't envision their pulling into the boat ramp area at the park any sooner than almost four hours from now.

"We'll stop on one of the sandbars for lunch."

Stop. For lunch. Stop.

As she put some muscle into her paddle strokes, she took comfort in the knowledge that a person couldn't perspire that much without losing at least a little excess weight.

A merciful breeze that started shortly after their picnic lunch on the sandbar helped their perspiration evaporate. Charlie seemed less pleased than she was.

"Could be a storm brewing."

"I know for sure that wasn't in the weathercast," Lucy said.

"Yeah, well, neither was this sauna."

Lucy slid her visor into her hairline to catch another drip.

———

Lucy could have kissed the cloud cover. What a relief! Between the breeze and the gray shade shielding them from the sun, and the traces of civilization showing up along the shoreline, Lucy thought they might actually make it all the way home. Her shoulder muscles ached. She only stopped to rub the soreness one time. That was enough to remind her they had no sunscreen.

She'd never been more grateful for current, the unseen that carried them forward even when they had to rest.

"Young people go canoeing," she said, aiming her words back over the lump of life jacket over her neck and shoulders.

"Yes, they do," Charlie answered.

"We're not young anymore."

"No, we are not."

Lucy reflected on all they'd seen along the river, despite the uncooperative heat. The beauty. The sounds. The surprise of birds. The wildflowers. Even the weeds seemed like art. Were art. A new scene around every curve. The beauty overrode the struggle.

And behind her the whole way, the man to whom she'd pledged the rest of her life. Different from her in an infinite number of ways. But the canoe kept moving forward.

A song. A song swelled in her heart as they made their final approach to the boat landing in Willowcrest. "Grow old along with me. The best is yet to be." Robert Browning would have appreciated how it sounded in her head.

"So you know the routine," Charlie said as they drew near to the finish line. "If you get out first, you can pull the canoe in farther. We'll ease into the shallow area by the—Oh, here comes the rain."

"Feels good, actually."

"Let's get this done. Take your paddle with you when you get out. Can you manage the strap on the cooler, too? Wow, it's really coming down."

"Got it."

"I'll hold us steady. Watch the— Lucy!"

The noises. The screeching noises. *Make them stop.* If she could hold her head... *Move, arms.* The pounding. Why didn't someone stop that infernal pounding? *Evan, is that you?*

Lights. No, no lights. Please. I beg you. No lights.

The light grew brighter. Someone pulled the sun right down in front of her face. Didn't people know that was danger— Dangerous? A person could go blind.

The sun dimmed a precious little, then went out.

Someone's choking me! Charlie, where are you? Someone's trying to choke me! I can't... can't swallow. Breathe. Can't move.

Her skin hurt. She lifted her hand to touch the skin on her face. Fire ants. She'd been... bitten by fire ants.

"Lucy?"

Stop shouting.

"Lucy, don't try to talk."

She could open her eyes only a slit with that much light aimed at her. Monet. Too much light and—she opened them wider—a Monet scrub top.

"Welcome back, Lucy."

228

She tried to answer. Nothing but a pained growl came out. The growl choked her.

"Try a sip of water. Just a small sip. No? That's okay. We'll try again later. I'm Meredith. I'm your nurse until seven. We'll let you get a little more adjusted and then let your family in for a few minutes."

Family. Charlie. The canoe. Slipping on the boat ramp. Concrete racing toward her.

"It's a good thing you were wearing your life vest, Lucy."

Did she drown? No. What?

"You might have broken your neck in the fall, if it weren't for that padding behind your head. You are one blessed woman."

Meredith dabbed at the corner of Lucy's eyes. The saltwater stung. Oh. Tears. Why did her face hurt so much? And the top of her head? And the lining of her skull? And her arms, knees, throat, chin.

Lucy reached to touch her chin. It felt like gauze.

"You can probably get rid of that gauze tomorrow. We're protecting the stitches."

Stitches? Her tongue felt twice its normal size. Did she have a throat anymore? Had they removed it? She couldn't swallow. Couldn't talk.

Couldn't...think.

"I'm going to put more ointment on the scrapes on your knees. And it looks as if your sunburn could use something soothing. Am I right?"

Was she intubated? That must be it. *Please take this tube out so I can swallow and talk.* Lucy lifted her arm and signed, "Please," then pointed to her throat.

"You know sign language? That will help so much during your healing."

She couldn't move her head. Moving only her eyes let her see more of the room, but the action felt like knives pushing her eyeballs.

"I should save a little of the sunburn salve for your husband, huh?"

Charlie.

"This is just aloe vera gel. I think your doctor is considering ordering a corticosteroid cream if it gets worse. But, we have a few other things to take into consideration, don't we?"

The gel felt cool. Somehow that helped with the headache. Fierce headache.

"We've got some good pain medicine pumping in you now, Lucy. You should feel it kicking in soon. That last IV site didn't help at all, did it? I'm going to do your knees now. This won't feel great at first, but try to hold still, okay?"

Can't move. Holding still won't be a— Ouch! Stop. Stop it!

"Only a little more. Can't let these scrapes get infected."

The woman wasn't yelling at her. All sound seemed magnified. The incessant beeping of the IV machine. That other machine with the green lights.

When Meredith pulled the sheet over her knees again, Lucy gestured for something to write with.

"Sure. We have a white board and markers for you. They'll be your best friends for a while." She held the small white board and put an uncapped marker in Lucy's hand.

Uncapped. A small kindness that made the tears flow again.

Lucy wrote, "Please take tube out of throat. Can't swallow."

Nurse Meredith read what she wrote and pressed her lips together. Her eyes showed a sympathy that both warmed and frightened Lucy.

"Hon... you don't have a tube in your throat. That's all swelling from your accident."

A half-dead person trying to draw a breath. That's what Lucy's voice sounded like when she tried to talk. Chin, jaw, teeth... what didn't hurt? Her hands. She'd been carrying things. The paddle.

The cooler. Her water bottle. That's why she didn't brace her fall with her hands. The underside of her forearms were a mess of raw flesh. It hurt to lay her arms on the hospital sheet. But it hurt to lay them anywhere else, too.

"Lucy?"

Charlie.

"LucyMyLight, I'm so sorry."

Why was he sorry? She was the one who slipped. She should have been more careful. He always told her to be careful.

She pointed to the lower half of his face and arms.

"Cute, isn't it? I call this line across my nose the equator. If my hat had a little deeper brim..."

His smile didn't fool her. She could tell he had a hard time seeing her like this. His eyes kept darting from her chin to her arms to the noise-making machines.

"You took quite a hit." His eyes teared up. "I'm just glad you're..."

Alive? The medicine must be kicking in. She was more grateful for that than she had been a half hour earlier.

Maybe the pain medicine would allow her to swallow. She tried. Her saliva—what little of it there was—got stuck at the barricade of her swollen-on-the-inside throat. She choked again, grotesque sounds coming from her mouth.

"Lucy," Meredith said, leaning over her. "Breathe slowly. Let it pass. You're okay. You're okay."

Sweet face, that Meredith. Not much for brains, though. Lucy was a million miles from okay.

30

It helped her throat when she sat up a little more. But it made the headache worse.

I tripped and fell. Up an incline. How did I wind up here rather than at home with an ice pack or two and ibuprofen? Talk about your romantic dates gone awry...

Charlie would have applauded that her sense of humor came back. If he could read her mind. And if she could smile without her face bones hurting so much. Each shift's angels of mercy worked hard to keep her from catching more than a glimpse of herself in the mirror. Even when they took her to the bathroom a few steps away from the hospital bed, they positioned themselves in front of the mirror. As if Lucy could have raised her head to look up.

Torture, those trips. She held it as long as she could to minimize that necessity. Her knees rebelled, but it was the pain in her head and neck that ratcheted past ten on the medical pain scale when she had to be upright for longer than a minute or two.

Lucy kept a washcloth balled in her hand to dab at the saliva that dribbled out of the corners of her mouth. After two days with the aloe vera, her sunburn had removed itself from the list of priorities. One thing: check.

She hadn't shattered her kneecaps, just scraped the flesh from them. Two things: check. Well, three, if you counted each knee separately. At this point, scrambling for positives seemed perfectly reasonable.

No broken bones. None. Surface injuries, except for the concussion and the damage in her throat from the sudden jarring and the hyperextension when she caught the bulk of the impact right under her chin.

They talked about sending her home. Home. She couldn't imagine.

First, she had to relearn how to swallow.

The specialists weren't all that concerned about the fact that she couldn't talk. They said it was possible—possible—she'd regain that ability when the traumatic swelling receded. Charlie—who rarely left her machine-, equipment-, and people-clogged room—had brightened. "So she'll get her voice back?"

It didn't escape Lucy's notice that the medical staff paused a second too long before answering, "The ability to talk, Mr. Tuttle. Her vocal cords were stretched beyond their limits. We have no way of knowing what her speaking voice will sound like if it does return."

"And singing? She's a music teacher."

Was, Charlie. I was a music teacher.

"Are you a praying man?"

Charlie had nodded.

The doctors switched subjects and addressed Lucy. "Your primary assignment right now is to master swallowing. We need to get you off the IVs and graduated to at least clear liquids before we can send you home. Pain meds? We can do those through patches or liquid versions if we have to. But you need to be able to swallow, Lucy."

I'm not fighting you on that idea, just so you know.

"To be up front with you, another forty-eight hours without swallowing and we'll have to consider a feeding tube directly into your stomach, and we all want to avoid that if we can."

A swallow therapist—who knew they existed?—would work with her. The concussion would have to heal on its own.

She started crying before the swallow therapist said anything beyond his name. He'd been one of her problem students more than a decade ago. A cutup in class. Never practiced. Somehow managed to pull it together for performances and contests. Loads of talent. Rich baritone voice. Not much discipline.

How many humiliations would this life hand her? She probably looked like a motorcyclist with the worst case of road rash, wearing a hospital gown that had been washed a dozen times too often—the design, faded to indistinct swirls—unable to smile or lift her head without grimacing, which brought more pain. *Hey, Brogan! Good to see you! Tell me all about your life since high school. Me? Oh, I'm doing great. Just great. Retirement is more than I could have hoped for.*

Instead, she lay there like a helpless, blubbering, pathetic lump of barely recognizable humanity, having to swipe at her saliva and her tears with a washcloth.

"So," he said, "you came in for a chin lift and the surgeon got a little scalpel happy?"

Same old Brogan. She grunted.

His expression sobered. "Seriously, Mrs. Tuttle, I'm so sorry about what happened to you."

She touched the front of her throat gingerly.

"That, too. I meant getting fired."

She widened her eyes to correct his wording.

"Having the program dumped. Not cool. What you and music did for me..." He looked away, then back. "I don't know where I would be today without your influence."

Brogan? The bane of her existence, Brogan? Charming, but not-living-up-to-your-potential Brogan Meyers?

"Did you know that I use a lot of the vocal training techniques you taught us when working with my patients?"

Wonder of wonders. He'd retained something even when it seemed he hadn't been paying attention to anything but the girls in chorus and show choir.

"They call me the Swallow Coach. But my real title is Certified Speech and Language Pathologist. I work with dysphagia patients all the time. Most of them are older than you are."

She was unusually young. That qualified as a positive.

"Can you show me your range of speech and swallowing ability right now, so I know where to start?" He drew closer, watching her eyes and throat, alternately. "Any minute now. Any. Minute."

She motioned for her whiteboard and marker.

He smiled. "These are not crutches you'll be allowed to use for long, Mrs. Tuttle. We're going to get this healing process moving." But he handed her the board.

She wrote, "I choke when I try to swallow. No room in there. Plenty of pain, though. And by the way," she added, "it's nice to see you again. So proud of you."

His blush was worth waiting more than a decade to see. "I...I may let you keep the whiteboard a little longer."

She reached to wipe the message from the board with the felt eraser clipped to its side.

"Wait!" he said. "Do you mind if I...?" He pulled his cell phone from his lab coat pocket. "You can erase the pain part. But I'd love to send the 'proud of you' part to my mom. She would get such a charge out of that."

She smiled. It hurt less than it would have minutes earlier.

"Oh. No. Sorry. Privacy issues." He shook his head. "Sometimes regulations just plain get in the way. Not that everybody in the whole town doesn't know you're in here."

Lucy erased the first part of the message, then took the marker and signed her name with a flourish underneath "proud of you." She held out her hand and closed her fingers quickly to indicate she wanted the phone. He held the whiteboard in front of his chest so she could snap the picture.

"That," he said, "was a moment. Thank you."

Simple as it was, the small message pressed to his chest did something for him. His presence in the room did something for her. She'd learn to swallow, if only for him.

Charlie entered the room with a sack of fast food. The smell of the food beat him into the room.

"Mr. Tuttle?" Brogan said, setting the whiteboard aside. "I'm Brogan Meyers, your wife's star student"—he winked in Lucy's direction—"and now part of her medical team. We're working on some muscle training exercises."

"Great. Thank you for all you guys are doing." Charlie set the bag on the window ledge near the chair he'd claimed as his own. A recliner, no less.

"If I could ask you to please step out of the room until we're done with our session? We've found that having an audience isn't conducive to...um...full patient cooperation."

"Oh, sure. Sure, no problem," Charlie said. He patted Lucy on her nearest foot. "I'll be down in the visitor lounge at the end of the hall. Call if you need...I'll check back later."

He started toward the hall. Brogan grabbed the bag of food and handed it to him. "Food smells don't help either, so—"

"Right. Right. I should have thought of that."

"It's all a learning curve right now, Mr. Tuttle. I'll come and get you when we're done."

When Brogan returned to her bedside, he said, "First order of business. While you're healing, you need to pull up your big girl—" He halted himself before she had to raise her teacher stop sign hand. "What I meant was that it's going to take a ton of courage. Not just with the exercises. They won't feel good. Bottom line. But with expressing what you really need. If it's privacy, then say it. If it's telling your husband that the smell of food makes you nauseated or tightens your throat even more, tell him."

He adjusted the cooling pillow under her head. "People won't know unless you tell them."

She managed a small nod while signing *yes*.

"ASL? What do you know? I always thought you directed with...with elegance or something. I didn't know you signed."

"Learning," she whispered. More mouthed than whispered. Her throat responded by clamping tighter.

He stumbled backward in a theatrical move with his hand over his heart. "I heard that, Mrs. Tuttle. Your first word since the accident?"

She signed *yes*. Her first word. *Learning*.

"Do you sign for worship services at your church?"

He'd remembered faith mattered to her. She signed *no*. She didn't know enough, plus it hadn't occurred to her, plus she hadn't been in a worshipful mood lately.

The thought clamped down hard on her heart. A new source of pain. Maybe she did have broken ribs. No, the life preserver had protected her from that, too. The Life Preserver. If she ever got a moment alone, a moment when her head didn't threaten to fall off, she needed to spend some time thanking her Life Preserver.

"Your first real word. That's a good sign. Okay, let's get started on some simple, low-stress exercises."

Nothing that happened within hospital walls could be labeled low stress, in Lucy's book. Including what Brogan deemed simple exercises. They exhausted her and made her well of hope feel like a dusty California reservoir.

"Don't worry about not sensing any improvement right now, Mrs. Tuttle. We're building to that. Time and the treatments you're getting are the only things that will help that swelling go down. Until then, we're going to be satisfied with baby steps. Okay?"

Brogan taught well. His instructions were clear, expressed with word pictures anyone could understand. And flawlessly patient even when he had to be firm. Keeping the tears of gratitude at bay might be her most formidable challenge.

"Your concussion is a barrier. I won't lie to you. Not that I ever did," he added, eyebrows angled like they used to when he tried to excuse his way out of detention. "I'd have you doing more neck stretching and posture exercises if it didn't make your skull scream."

How did he know?

"I'm leaving this chart of the techniques we learned today. Tomorrow," he announced with flair, "we'll learn the Mendelssohn maneuver, with your primary care's permission. And no, it has nothing to do with Felix, the composer. Different spelling."

He remembered her classroom assignments about Felix Mendelssohn? Seriously?

"Although," he said, "maybe there is a connection." He stood back, his electronic tablet tucked under his arm. "Did you know—well, sure you did—that Mendelssohn, Felix or Fee as I like to call him, advocated for the revival of Bach's *vocal* music in the nineteenth century? Ironic, isn't it?"

She snapped her fingers for the whiteboard.

"Ask nicely, young lady." He assumed a posture much like the one he must have seen her use hundreds of times in his school career.

She grunted a chuckle and signed *please?* She wrote, "Mendelssohn? Really? Are you now a fan?"

"I study Renaissance men in my spare time. And a few other things. I'm on the trivia team for my alma mater alumni reunions."

Oh, good grief. Proof positive miracles still happen.

31

Psalm 104:33: "I will sing to the Lord as long as I live; / I will sing praises to my God while I'm still alive."

She reread the message on the small garden stake in the massive azalea plant Charlie bought her. It spilled over the bedside stand. The nurses all commented on how beautiful it was then surreptitiously nudged it enough to allow them to use the stand top for scissors, IV tape, gauze, ointments...

Reading the message meant two things. Her eyes focused better now. And she could roll her head a few degrees to the right or left.

Three things. Healing her soul had a ways to go. The garden stake plunged deep into the part of her that had yet to reconcile the way life had turned out, and the viable threat that she might never, would probably never sing again.

What kind of life was that?

As if cued, Sasha poked her head around the door to Lucy's room. She tapped lightly. Lucy waved her in, Sam on her heels.

Sam had visited earlier, including the time before Lucy had fully regained consciousness. But this was Sasha's first introduction to what a mess looked like. Her fine-featured face—so similar to Olivia's, worked hard to maintain its composure. Lucy could see it in the way her jaw flexed and the corners of her mouth twitched.

Lucy wiped drool from her chin, then held her gauzed arms out to Sasha. And just where was Sasha supposed to hug her? Lucy dropped her arms and patted a place on the bed. Sasha eased onto the spot.

Bless you, child. You'd be amazed how many people, including your father-in-law, don't realize that clunking against the bed jars this bruised brain of mine. Lucy would tell her, if she could.

Charlie and Sam carried on a hushed conversation in the corner by Charlie's chair. Sasha sat for a long while holding the tips of Lucy's fingers. She reached for the bottle of hospital-issue lotion and held it questioningly. Lucy nodded.

Sasha warmed the lotion in her palms, then massaged it into Lucy's neglected hands. Around the cuticles. Each finger. She took care to avoid the IV site, but tended the dent where her wedding ring belonged. The emergency room crew had removed her rings, in case her hands swelled. They hadn't. But hospital policy recommended leaving them off until she was at least finished with IV fluids.

One more reason to relearn how to swallow.

Sasha kept her focus on Lucy's fingers. Then she lifted her featherweight off the edge of the bed. She freed the sheet from the bottom of the bed, folded it back, and massaged Lucy's feet.

Lucy fought to keep her throat from convulsing. She practiced the deep yawn Brogan suggested. She didn't even try to stop the tears that trickled to the pillow behind her head.

Sam asked, "Does that hurt, Mom?"

She signed and mouthed *no*.

"Do you need her to stop?"

No. *No.*

"Is there anything I can do for you?"

She signaled for the whiteboard. "Keep appreciating that girl." She drew an arrow pointing toward Sasha. And added an exclamation point.

Sam choked up. He nodded his head, but didn't speak. There was a lot of that going around. A veritable epidemic.

"Son," Charlie said, "do you want to buy your old man a cup of coffee? Leave these two alone for a bit?"

Sam's hesitation was palpable. But Sasha nodded to him and to the whiteboard, as if to say, "We can talk without an interpreter."

But they didn't *talk* for many minutes. Sasha finished her massage, draped the sheet over Lucy's feet, and replaced the bottle of lotion. Lucy handed her the waterless hand cleaner from the over-the-bed tray. The sight of it made Sasha smile. She smeared some on her hands and held them palms out and over for inspection.

Lucy asked for the whiteboard. "Sasha, can you talk?" She'd wondered if the woman was unable or if she hated the sound of her own voice, as Lucy now did...if you could call one word a voice.

Sasha's brows dipped momentarily. "I...can...talk," she said, the sound thick, rumbling, indelicate, throaty, robot-like. Completely unlike the woman from whom the words emanated. And yet, beautiful to Lucy's ears. "Too...slow," Sasha added.

Lucy wrote, "I understand. More than ever before."

Sasha took the marker and board and wrote, "Pain?"

Lucy cupped a hand over her throat then stretched both hands inches away from either side of her head and moved them in and out, throbbing.

The pale yellow sundress Sasha wore floated around her when she walked to the window, her back turned. She stayed that way only a moment, then returned to the bedside, signing something dancelike, ethereal, but lost on Lucy. Or was it?

She caught a word she knew, then another. *Lie down. Water. Quiet. Soul. Path. Valley. With.* No, *with me. Table. Head. Cup. Life. House. Lord. Forever.*

Even without voice, the words held power to infuse hope.

Lucy pointed toward the azalea. To the garden stake. Sasha leaned down and read it. She faced Lucy. Lucy watched her lips and hands. *Sing. Live. Praises. God. Alive.*

Alive. The hand gesture Lucy had used with her students to get them to breathe from the bottom of their lungs and sing from their hearts.

"Nice nap?" Charlie kissed her lightly on the unsunburned stretch of her forehead.

"Drugs," she whispered.

"What?" He leaned closer.

Could she say it any louder? No. She waved him off and pointed to her IV site.

"Ah. Medication."

The world's worst lip-reader. She patted his hand. But he was all hers.

What day was it? She had an appointment with Dr. Hanley coming up. And the HHATT club must have met without her. Or not. Maybe it only felt as if she'd been hospitalized the full length of a prison term.

"More flowers, LucyMyLight."

She turned toward the deep windowsill. An embarrassment of riches.

"Who?" she asked, her voice still coming out with the grace of an ogre with a head cold.

"Seven new ones." He smiled.

She reached for the whiteboard and wrote, "Who are they from?"

"Oh. A bunch of different people. Former students. People from church. One's from, let's see"—he pulled a small card from a clear plastic pick stuck in an *avant garde* arrangement of flowers in a low fuchsia container—"'Carole, Angeline, and Marta.' Your book club, right?"

She nodded.

"They say, 'Missed you. Saved some hope for you.' Wait. Nope. That's what it says. Thought it was a typo."

Such surprising friends. Cherished.

"A whole stack of cards we haven't read, too."

Blessed. When she felt better, she'd start on thank-you notes.

Her lips—so dry. She reached for the tray drawer.

"I can get that for you." Charlie beat her to the drawer. "What do you need?"

Brogan's counsel revisited the room. *Tell them what you need.*

She pulled the whiteboard between them. "I need to be able to get my own lip balm."

Charlie's small pout quickly faded. "Got it." He stepped back a foot. When she retrieved the lip balm and applied it, Charlie gave her two thumbs up and a goofy grin.

Recuperating at home would be an adventure.

"Olivia said she'd be up to see you later."

"It's later." Lucy's daughter—joy of her heart—didn't bring flowers. She arrived with an arsenal of chocolate. "Swallow incentive," she said. "And it keeps well."

Lucy knew what that meant. If she couldn't get her act together, couldn't force her throat to cooperate, if she had to resort to a feeding tube, if it were months before she regained that simplest of body functions—swallowing—the chocolate would still be there, calling to her. Egging her on.

Olivia set the gift basket of chocolate on the tray table and bent to kiss Lucy on the— She paused. The young woman searched her mother's face. Lucy pointed to a spot on her cheek that no longer stung or ached. *Improvement. Somebody mark that down on my chart.*

"So," Olivia said, "any news?"

Lucy signed *no* while Charlie said, "Making progress. She looks good, doesn't she?"

Charlie, that's what you say at a funeral home.

Lucy pointed to Olivia. *You?*

"Maybe."

What was the sign for Lucy to say, "Spill it, girl!"?

Olivia glanced at the door. "Remember that date I gave up when you decided to eat concrete?"

Nice, Olivia. Her date with the onion ring guy. Another thing Lucy had fouled up when she fell. That heartbroken girl.

"I brought him with me tonight."

No. No, no, no, no.

"Don't worry, Mom. He's not the kind of man who cares what you look like."

Lucy squinted, begging for clarification.

"Not what I meant. He's a quality guy." She leaned in to whisper. "He insisted we stop here before we go to the airport."

Lucy's expression must have scared her. "Mom, no. Not me. He's flying out to California, for a gig. I'm seeing him off. We're having dinner together first."

Lucy grabbed the whiteboard. "I can barely hold my head up."

"Then don't," she said. She rearranged the tangle of curls that had once been Lucy's hairdo. Gently. Until now, even the air conditioner breeze made her scalp hurt. Olivia tucked her mother's hospital sheet a little higher under her arms. She leaned to Lucy's ear. "I really want you to meet him. And for him to meet you."

"Like this?" Her voice would scare small children into years of therapy.

Olivia smiled. "Just as you are."

Lucy lifted her hands and dropped them to the bedcovers.

"Great! I'll go get him."

She was back in ten seconds. He must have been right outside the door. Listening.

"Mom, Dad, this is Caden Cole. Caden, my mom and dad."

Caden. Like cadence, only not. Nice name for a musician.

If it hadn't looked so right on him, and if he'd been her son rather than her daughter's new boyfriend, Lucy would have brushed the long, thin curls from his forehead. And asked him to shave. But somehow...

A guitarist if there ever was one. A guitarist with Josh Groban hair after two months on a deserted island with only clam shells to shave with and an ample supply of shampoo.

Caden Cole dressed smartly, for a gig musician. Dress slacks, a V-neck tee, but in what looked like a high-quality fabric. Must have been the "classical" part of classical guitarist that influenced his hygiene and continental look.

Charlie stepped forward and took his hand. "Nice to meet you, Cade."

"Caden, Daddy."

"Mr. Tuttle, a pleasure. We're... we're praying for you, man."

Olivia took his arm and drew him toward the hospital bed.

My breath must smell like a compost heap.

"Caden, my mother." She stepped back then so he could come closer.

"I have such admiration for you, Mrs. Tuttle." He held her hand with a light touch. She could feel the calluses on his fingertips. Professional calluses.

"She's doing just fine. Aren't you, Lucy?" Charlie put his arm around Olivia. "She's going to be just fine."

Caden stole a look at Charlie. "For her courage here, yes. But I meant in her career as a music instructor." His eyes bore into hers. Olivia was right. Eyes worth looking at. They held compassion and integrity and a knowingness that put her at ease despite her circumstances.

"Both you and your father," he said.

She knew this boy. A former student? Maybe. Young men did have a tendency to change from their elementary years to their late twenties, early thirties. How did he know her father? Her brain couldn't think that hard yet. It might have simply been that Olivia had described him so well in their late night chat. No. No. She breathed evenly. Think. Mrs. Schindler hadn't moved to town until a handful of years ago. Caden Cole would have grown up in some other community.

So dry. Her mouth was so desperately dry.

"You know about what Lucy did? Does?"

Did, Charlie.

Caden's smile matched the warmth of his eyes. "I do." Then his gaze drifted to the IV stand or whatever caught his attention in the corner beyond it. "I've heard so much about the music program at Willowcrest. And not all of the information came from my mother."

No broken ribs. Confirmed on the X-rays. Then why did it feel as if her chest were collapsing into her spine?

"And"—he looked at Lucy again, his face twisted with what looked like concern—"you and this whole community deserve an apology for what she's done."

32

Who smeared the little dots in the ceiling tiles?

A wave of nausea smacked her. She pushed her call button.

"What is it, Mom? What can I get you?"

Lucy pushed the button again. *What's taking them so long?*

"Mom, you're scaring us. What's going on?"

"Yes? May I help you?" The unit clerk at the desk didn't even question how Lucy was going to answer her? Olivia stepped in.

"Something's wrong. Mom looks... I don't know... pale and— Dad, don't shake her. The concussion?"

"I'll send someone right down." The speaker near Lucy's ear stopped buzzing.

Meredith. She cleared the room of humans and leaned close, her stethoscope over Lucy's heart. Lucy waved at the whiteboard. "Going to throw up."

"No, no. Can't have that. Your antinausea medicine is supposed to take care of that. We don't want you to aspirate, hon. With all that swelling... Okay. Deep breath."

Really? Really?

"Slow and deep." While she talked, Meredith grabbed a clean washcloth from the stack in the side table drawer and poured from the unused, never used pitcher of ice water to wet the cloth. She squeezed the excess into the wastebasket with one hand while stroking Lucy's shoulder. "Slow and deep." She laid the cool cloth on Lucy's forehead.

Lucy's intake of air sounded like the noise fourth grade boys make when they blow through a blade of grass gripped between their palms.

Meredith dashed around the foot of the bed and flipped on the small fan Lucy was sure they'd never need. "I'm going to raise the head of your bed a little bit more, Lucy."

Lucy closed her eyes. The room still swirled. She opened them. *Something, some focal point. Orient. Stabilize. Seasickness technique. The clock. No. Moving hands. Meredith. Her face.*

"Keep breathing nice and slowly, Lucy. Slow it down. Good. Let me see that IV. Oh, hon, it's infiltrated again." Meredith pressed the call button and flipped the cloth on Lucy's forehead. "Need help in 218. Send Lois with an IV kit. STAT."

Lucy felt until she found Meredith's hand. She squeezed, then eased up as the wave inched its way off stage and finally passed into blessed oblivion.

"A little better?"

Lucy didn't dare nod, but signed *yes* with the hand not clutching Meredith's.

When the second nurse arrived, Lucy heard voices in the hall. "Is she alright? What's going on?" And then another voice—Lois?—instructed Lucy's family to give them space to work, to wait in the family lounge.

I'm okay now. It's passed. She managed a smile.

Meredith flipped the washcloth again. "A little scary, wasn't it?"

Smallest of nods.

"Lois is going to restart your IV. We'll get that medicine doing its job more efficiently again. And you'll be back on track for making more progress. Your forehead's still a mess of creases, Lucy. Just rest. That's the best thing at a time like this."

Brogan arrived just as Lois left. "Bad timing?"

"We had a little excitement for a few minutes," Meredith said. "Better now. I'll leave that to you and Lucy to decide whether she's ready for you."

"Hey, Lucy! You're sitting up more today. Good to see that."

"A bit of a necessity," Meredith said. "Didn't want her to aspirate."

"Oh," he said, a gentle hand on Lucy's foot. "That will remain a concern until we get that swelling down and those muscles and ligaments retrained." He observed her now-quiet breathing. She did her best to prove she could do it—breathe quietly.

Meredith unclipped the pager from her scrub top pocket and glanced at the screen. "Are you okay if I leave you in Mr. Meyers's capable hands?"

Lucy nodded. A real nod. The ceiling dots remained dots.

She wrote on the whiteboard, "Family worried. Lounge."

"Gotcha. I'll get someone at the nurses' station to let them know it'll be another few minutes, but that you're okay."

She signed, "Thank you."

He was gone less than a minute. She spent the whole time trying to think of anything but the exercises Brogan would put her through. She missed Evan, his bright blue eyes, the way his hair fell over his forehead as if styled to look casual, the feel of his small hand in hers as he depressed the piano keys and discovered he could make music, his response to "Jesus Loves Me."

She'd been singing.

Was that the last time? Ever? No. She'd sung as the canoe approached the boat ramp, hadn't she? What an inglorious finale.

After three or four failed attempts at the simplest of the exercises, Brogan pronounced her sufficiently traumatized and postponed the therapy session until the following day... during which she'd reach the dreaded forty-eight hour mark. He promised to come before that ominous hour for one last try. She promised to keep working on it.

"Okay to come back in?"

Charlie. She turned her head and motioned him in.

"You look a sight better than you did a bit ago."

She wrote out, "Compliment?"

"Had me worried."

She took his hand and held it to her cheek. It couldn't make the ache in her wrenched jaw any better, but it soothed the deep ache in her soul.

He pulled a visitor chair next to her and rested his elbows on the bed. "They got rid of the gauze on your chin, huh?"

She guessed so. Had Meredith done that?

"Want me to count the stitches?"

Who said they'd been married so long they had nothing to talk about?

She rested her hand on his shoulder. Strong, sturdy shoulder. Always there. Always. Sam's body type was more like her father's. *Daddy, I'm glad you didn't live to see all this.*

"Olivia?" Lucy knew ninety-year-old men with smoother voices than hers.

"They had to take off or he'd miss his flight. They missed their dinner, but I have a feeling he'll be back. Nice guy, don't you think? Despite his...parentage?" He watched her face.

No, Charlie. I'm not going to have a seizure. It was a complete coincidence that I thought I was going to— How can this man need a haircut so often? Those silver tufts.

"Knock, knock. Mind if I barge in on this love fest of yours?"

Charlie leaned back. "We have the rest of our lives, Meredith. What do you need?"

"Have this idea pestering me. The only way to get rid of it is to face it head on. You don't have any allergies listed on your chart, Lucy."

She nodded.

"Nothing."

No.

"Did you ever take anything or eat anything that you thought, 'This just isn't sitting right,' or had any kind of reaction at all?"

No.

Charlie propped his hands on the bed. "What are you thinking? That's she's allergic to something you're giving her?"

"Don't know. Just a hunch. But there are a lot of medications or food allergies that make people feel as if their throat is closing in on them. Couple that with the damage done in the fall..."

"Think, Lucy. Anything?"

She shrugged. Nothing that she could remember...

"What's that look mean, LucyMyLight? Here. Write it out."

"OTC cold medicine. Tried once. Didn't like how it made me feel." She touched her throat and wrote, "Clogged. Not right."

"Which one? Can you remember?"

She wrote the name on the board.

Meredith shook her head. "I don't think there's any crossover with what we have you on. But I'll check to make sure. When was this you tried the over-the-counter med? Just the once?"

"Found something that worked better. Once. Maybe five years ago."

"Give me a little time to investigate. I'll get back to you as soon as I see if there could be a connection."

"I like her," Charlie said.

Lucy pointed to herself and held up two fingers.

"You sense victory?"

Oh, Charlie.

"Me, too, Lucy."

Lucy woke Charlie by banging on her tray table when he started snoring in the recliner. She sent him home, telling him she'd rest better—and she would.

Her new IV was in an awkward place too near the crook in her elbow, she thought. Although the nurses had helped reposition her in minor ways since she'd been admitted, the pain in her head had kept her from lying on her side. This night, that's all she wanted to do—turn on her side. Between the IV and the throbbing ache in her jaw—how was it not broken?—that wasn't about to happen. She rotated one ankle, then the other. Her knees itched. They must be healing.

The most recent trip to the bathroom had been easier on them all. She hadn't had to lean so heavily on the aide who helped her, had been able to raise her eyes to the mirror. Not her smartest move, but she still considered it a victory.

"Still not asleep?" Meredith kept her flashlight pointed at the floor as she checked Lucy's IV site and the infusion machines.

"You still here?" Lucy growled.

"Covering for my replacement who had a fender bender tonight. She'll be a little late. Dr. Scott wants us to try weaning you off one of the medications. He listened to my allergy theory and couldn't find any connection with what you're on. But he wants to get rid of this one eventually anyway. Baby steps."

Lucy nodded.

"Do you want something to help you sleep? You seem more restless than usual."

Was it courageous to say no, or to admit she needed help?

"I'll take that as a yes. The prescription's waiting there for you if you need it. There's no shame in taking help when it's offered. You'll heal better if you sleep better."

Lucy signed, "Thank you."

Sleep. Lie down. Stilled waters.

Hadn't she sent Charlie home hours ago? Her eyes drifted shut again. Open. Not Charlie. Olivia. Bless her heart.

How had she gotten onto her side? She rolled back, but a pillow blocked her.

"Good morning, Mama."

The term of endearment Olivia reserved for the tender moments. Lucy waved good morning.

"Do you need that pillow out of the way?"

She nodded. Was there more range of motion? She tested, dropping her chin a little, raising it beyond ninety degrees. Her stitches pulled. Her neck rebelled. But she had more movement.

Lucy pulled out the tray drawer while Olivia watered the flowers and plants clogging the windowsill. Lip balm. But her mouth wasn't as dust dry as it had been.

She tapped twice on the tray table.

"What is it, Mom?"

"Water." Still a groggy ogre growl.

"Yes. I think I got them all. Oh, the azalea."

She tapped the table again. "Water, please." She pointed to the plastic covered water jug so faithfully dumped and refilled every shift.

"Are you sure? Can you do that? Want me to get the nurse?" Olivia reached for the call button tucked among the pillows.

"No. I try."

"Okay, but I'm keeping my finger on the button."

She grabbed a small waxed paper cup from the dispenser above the sink and poured a dribble of water into the cup. "Mom, this is lukewarm. I'll get you some fresh ice water."

"Please."

Olivia made a show of her finger hovering over the call button while she handed Lucy the small cup.

She relaxed her throat, focused on loosening what had remained locked tense since the accident. The water on her lips would have been enough. Maybe she should stop there. Doubts swirled in the waxed cup.

But she tipped the cup and let the teaspoon of water into her mouth.

Olivia held a washcloth in one hand and the call button in the other.

O ye of little—
She swallowed it!
"Mom, did it go down?"
She nodded.

Olivia dropped what she'd been holding and danced—arms flailing—over every inch of tile in the room. "She did it. She did it. My mama she done did it. She did it. She did it. My mama she done did it!"

It hurt like fire, but she did it. "Hurt like fire" didn't fit well into the rhythm Olivia had going, so Lucy didn't mention it.

33

"Show off," Brogan said when Lucy demonstrated what she'd accomplished. "Now what do I have left to teach you?" He faked a dramatic checkmark on the ever-present tablet he carried.

Lucy kept the whiteboard in bed with her now. She answered him with a sentence written in red. "Help me learn how to swallow more than a sip of water." She smiled, but added, "And to talk again."

"Unfortunately," he said, leaning on the tray table, "my work doesn't come with guarantees. I wish it could. But look at the progress you've made so far. High five. Fingertips only. We don't want to jar that brilliant but battered brain of yours."

Yes. She would definitely write that boy's parents a letter.

From her position in the corner, in Charlie's chair, Olivia offered two thumbs up. Just like her dad. Lucy used her left thumb to tell her daughter to take a small hike.

"I'm going. I'm going," she said as she grabbed the novel she'd been reading and slid into her flip-flops. "I'm going to the courtyard. Somehow, I'm going to get some sun before this summer is over. Oh, apologies to your face, Mom."

Every mother's dream compliment. The sunburn wasn't an issue any more. That left Lucy with only a dozen true issues.

"I could read most of it on your face," Brogan said after Olivia left them alone for the therapy session, "but what was your pain level when you swallowed? Give me a number."

She held up six fingers, then added a seventh.

"It's a start." He scrubbed his High Five fingers through his short-cropped blond-tipped hair. In high school, he'd worn it—she thought hard for a description—undisciplined. Brogan pulled on surgical gloves. "Mrs. Tuttle, I need to determine the mechanics you're using for swallowing right now. So I'm going to lay my fingers on your throat..."

She flinched.

"...very gently. Like this. Are you okay?"

She nodded, concentrating on relaxing her neck and throat as best she could.

"Now"—he spoke with the signature calm of a yoga instructor—"take another sip for me. Like you did before." He handed her the small cup.

She tipped it to dribble into her mouth—no straws allowed yet, he said—and worked on getting her swallow reflex to follow her mental directions. Tried again. She couldn't do it, with even that light pressure on the front of her neck. Her frustration must have registered clearly on her face.

Brogan stood back. "Okay, let's try this. You lay your fingers here, between your lovely chin enhancement stitches and where your Adam's apple would be if you were a guy. Use the back of your fingers. That's a more natural position for your hands. Now, when you swallow, pay attention to the pressure you feel against your fingers. I want you to describe it to me after you've swallowed."

But she couldn't. It's as if he told her to think about anything other than Hawaii, so of course that's all she could think about. Her throat. And how it wasn't moving. She'd lost the ability already.

After numerous attempts, she spit the dribble of water into her washcloth.

Brogan cupped his hand around his chin, then stuffed both hands into his pockets. "I don't want you to get discouraged. And yes, I know that's asking a lot. But you have so many things going on in there."

She rolled her eyes, although she would have sent him to detention if he'd done the same to her in the classroom.

"There's the hyperextension of the ligaments, your vocal cords, tendons, muscles. Larynx. Pharynx. There's the bruising of all that tissue in there from the impact. You've got your swelling and the body's natural resistance to pain. And, I see this all the time, swallowing comes so naturally..."

She turned her head to face the door to the hall.

"It comes so naturally that when we try to force our bodies to do it, we can't replicate the action. It's like telling your eye muscles not to twitch. What happens?"

Lucy turned back toward him and demonstrated eye twitching. Both eyes. Unsynchronized.

"Bingo." He laughed, then sobered. "Would it help you if I told you this could well take months? Don't frown at me. That's intended to give you hope that if you're not there today, cut yourself some slack. You haven't had a knee replacement, have you?"

Yesterday, mister, you told me I was too young. Was that yesterday? She signed *no*.

"Our orthopedic surgeons stress to their knee patients not to get discouraged if it takes as much as a year for the knee to feel like it's part of them. For the discomfort and swelling to disappear. Then, when it happens at seven months, they're jazzed."

She mimicked jazz hands.

"Yes! That's the attitude."

Brogan, they were sarcastic jazz hands. It was amazing to Lucy how much internal conversation took place when no one expected her to talk.

God, You expect me to talk someday, right? Sing again? The silence cramped her throat. *Say something.*

"Here's your assignment for today."

Brogan outlined another series of exercises, including a relaxation technique that reminded her of Lamaze classes in her childbearing years.

"I'll be back this afternoon for another session. We'll talk some more about the connection between music and language. That may interest you."

She attempted a smile.

"Music and speech have many common denominators," he said. "Think about it. Jaw placement, tongue placement, pitch, timing, timbre, tone. Your brain uses the same circuitry whether making music or language. Which is why," he said coyly, "music education is so critically important for the development of a more fully rounded person, and especially..."

She managed to raise her eyebrows at his fervor.

"...with those who need an educational boost in other areas or a developmental boost in language."

Sighing hurt very little, she was surprised to find.

Brogan's facial expression shifted gears. "I assume your doctors will want to keep you here a little longer despite your incredible victory swallows this morning."

Lucy groaned.

"And I believe they're planning another set of scans to see what's going on in there"—he pointed toward her head—"and there"—he pointed toward her throat. "Headaches any better? Dizziness abating?"

She nodded yes. Emotionless.

"Remember," he said, then jazzed-handed his way out of the room.

Charlie texted her shortly after Brogan left. "Not coming today. Down with bug. Evan has it too."

Olivia had a hair stylist appointment before long, then work. Sasha couldn't leave Evan. Sam was ninety miles away. God was a quadrillion.

No. She chided herself. Untrue. Near as her next breath. She hadn't been alone much since getting carted into this cramped, artificially sunny room. Processing thoughts had been impossible in the early days when brain activity screamed, "Cut it out!" She'd be alone in spurts today, between tests and vitals and therapy treatments and...

And what would that mean? She'd finally have to get honest with God about how her life felt like a pile of used worm bedding? As if He hadn't been watching.

He had. She knew He had. She came within centimeters, the doctors said, of a broken neck and complete paralysis. She could trace through the past few days and mark the vast difference between barely tolerating opening her eyes a slit and regaining her ability to read, between raging pain and a seven. How far would she be a week from now? A month from now?

She thought the job loss had silenced her. But she'd still had her voice then. Now she knew the full depth of silencing someone's song. Lucy held her hand to shield her from the glaring truth of the plant stake in the azalea.

Olivia stopped back in the room, kissed her mom goodbye, then headed for the stylist. Lucy warned her not to show up the next day with a cuter style than her mother sported. She pulled her fingers through tangled curls for emphasis.

The Teflon gauze on her forearms and knees was changed only once a day now. Those were the kinds of appointments that filled Lucy's calendar.

Most hospital patients beg to go home, don't they? When they start feeling better. Lucy sat on the edge of the bed, pushing herself up slowly from the side, as the nursing staff had taught her. It allowed her to face the window more directly. The outside world. Which was her problem. One of them.

Leaving here would mean stepping into a world she was pretty sure she wasn't ready for.

Sasha, how do you do it? How do you deal with people who don't understand what you're coping with internally? The Assumers. People who assume because you look okay on the outside—Lucy touched the stitches that would one day look like a simple crease—*you can and want to talk.* Or want to and can, in Lucy's case.

Leaving here would mean resuming her life. But none of it would be the same. Maybe for months. Maybe forever.

She'd wasted her grief.

The mourning she did when the RIF letter changed everything? Amateur stuff.

Her keys were in the cooler.

And her brain proved once again that it hadn't yet returned to linear thought. Life as she knew it was over, or in severe need of remodeling, and the thought she couldn't shake was her key fob in the padded lunch cooler at the bottom of the canoe.

Marked for demolition. Some crazy restoration contractor might look at her life and stand, fists on hips, in front of the bulldozer. "Stop! She can be saved!"

Uh huh. Who was going to believe that?

Believe. That.

Quit looking at me. The azalea had no response.

She'd watched enough HGTV to know it was possible. Wrecks of buildings not only made new but better than ever. Restore the hardwood floors. New cupboards and appliances. Get rid of the knob and tube wiring. New roof.

Lucy lightly patted the sun-scorched patch on top of her head.

She'd seen the befores and afters. *How did you turn that mess into that work of art?* Remarkable.

The work of an expert with an excess of determination, the right tools, an eye for design and function, and necessary resources. And time. And elbow grease.

Some people couldn't visualize the finished product. Homeowners interviewed early in the process stood an inch away from a drop-off into a crawl space of hopelessness. Lucy had always considered herself a visionary, more able than most to "hear" potential for music in the beginner instrumentalist's attempts or in the rapidly changing seventh grade boys' voices.

But this...this...

She poured a few drops of water into the sad little Dixie cup. Into the mouth. *Think of Hawaii.* She swallowed and gave herself two half-hearted thumbs up. One more time. Pour. Drink. Swallow. Pain level only a five or six this time.

The next time the nurse or aide came by, Lucy would ask for tea. Room temperature. Until then, she'd keep practicing. *Practice doesn't make perfect. It makes possible.*

Ten minutes upright at a time at first, her day nurse had said. "Working your way up slowly is so much more encouraging than backtracking because you overdid it," he warned. She obeyed, swinging her legs back into bed, lifting the sheet high to avoid sliding it against her knees.

Whoever invented the cooling pillow deserved a raise, she thought. Or an award. Genius idea. Her head and neck nestled into its embrace. Which medication had they removed? Not the one that made her sleepy. Ten minutes upright couldn't have made her this tired by itself. She closed her eyes. No point in fighting—

"Mom?"

"Sam?"

He leaned down as if to kiss her, but changed his mind. "I can't stay long. I'm on my way to Chippewa Falls for work. District meeting. Mom, it's good to hear your voice."

Lucy pointed to her throat and ground out, "*This* voice?"

Such tenderness in her son's eyes. "Any voice."

"I can swallow," she wrote on the board. "A little."

"I know," he said. "Sasha told me. Dad told her. Olivia told him."

Family.

"I brought your laptop."

Did she want it?

"I thought we could Skype. Especially now with half the household down with this cold. No fever. Just congestion. But you do not want it on top of..." He gestured with a wide circle encompassing all that was wrong lying on the hospital bed. "You've Skyped before."

She nodded yes, not having the energy to explain how she'd used it in her classrooms and with fellow music instructors across the country. Several had wanted to Skype over the summer. She'd found an ample list of excuses not to.

"And," he said, "it will give you a chance to e-mail, if you want to talk—communicate longer messages."

She wondered if she could reacclimatize to more than one- or two-word answers.

"I brought these, too. For your music." He pulled her Bose ear buds from his pocket.

SamWise.

Lucy thanked him.

"I have to run. Really. Maybe we can Skype tonight?"

She nodded. And held back her tears until he'd left the room.

34

Tea led to Jell-O. Jell-O led to warm bouillon. Bouillon led to more bouillon.

"Mashed potatoes?" she wrote on the board.

"Not just yet," Brogan told her. "So proud of you. But it's a texture issue. We have to take this in a careful progression. You're doing so well. You don't want to undo anything."

No. Even she felt the pout on her lips.

"If you keep making forward progress, I'll have to turn you over to an associate of mine who practices through the clinic."

Her eyebrows asked why.

"You'll be discharged soon. Won't it feel great to get home? To your own bed?"

My own bouillon. "Yes," she signed.

Brogan stuck one finger in the air. "And . . . that's another thing that's going to have to come to a stop. I know you *can* sign these simple words. But you can also say them."

"I sound grotesque," she said. The growl had advanced to low, airy hoarseness. Too weak for anyone standing more than a few feet away to hear.

He raised his eyebrows and asked, "How did you learn to swallow, Mrs. Tuttle?"

Zinger. "Practice."

"That's what it's going to take to strengthen your voice, too. But not in the way you'd think. We don't know how far we'll be able to bring it back."

"Please say farther than *this*." Her throat ached from the extra words.

He jazz-handed her into a smile. "That's our hope. Always hope. But with the kind of damage you have in your larynx, we need to rebuild muscle strength and train other muscles to support your vocal cords while they heal...as much as they're going to."

As much as they're going to.

"We'll start adding those vocal exercises to our swallowing routine. This will sound contradictory, but your larynx needs both exercise and rest. Straining it? Bad. Working it? Good."

Lucy didn't want to be transferred to another speech and language therapist. Didn't want to leave the house for appointments. Didn't really want to go out in public without a voice. Didn't want to put her focus on what might be a worthless pursuit. But she would. But she didn't want to. But she would.

"Family's not well," she said.

"Now, try that again, but intentionally lift your pitch. Don't let it growl down there in the basement. As you well know, that's hard on your vocal cords. We have to tell them not to be lazy, sluggish. Try again," he said.

"They're not well. My people."

"Short sentences, remember. A little at a time."

She sat up too abruptly. She grabbed at the tray to steady herself. But the room settled, like a snow globe toward the end of its blizzard.

"Are you okay?" Brogan dipped his head to look her in the eye.

Lucy nodded. Slowly. She picked up the whiteboard. "Docs can't send me home yet. Family sick. Summer colds."

"Could be a dilemma for us," Brogan affirmed. "A virus right now wouldn't necessarily kill you. But it would feel like it. Any place in your house you can kind of hole up and keep away from the others? Any room where you can lock yourself away? Some place you can retreat and have minimal contact?"

No. *Believe me, I've looked.*

Her office couldn't even hold a twin bed.

"Get thee to a nunnery?"

She shook her head.

"Hotel? Family friend? I don't recommend you being completely alone for a while. You're a choking risk, still. Each new food introduced is an experiment." He scratched the blond stubble on his chin. "I don't know what Dr. Scott will say. If I were you, I'd definitely talk to him about it, though."

She signed *yes*, then shook her hand as if erasing the sign and spoke the word in a bass rather than a contrabass voice. "Yes. I will."

"Good. Maybe you could stay with your daughter."

She wrote, "Olivia lives with us right now."

"Your son? Sam was in my graduating class."

"Sam lives with us right now."

"Any more children you could tap into?"

She laughed. It hurt, but she didn't try to stop it.

"Pitch a tent in the backyard? Okay, okay. I'll stop suggesting the impossible. It'll all work out."

Does it? Does it always all work out? Sometimes people lose everything.

"*You* were the one who taught me that it always works out," Brogan said, scribbling something on the updated prescription of her strengthening exercises.

She hadn't said that aloud, had she?

"Maybe not like we plan, or how we thought it would look." He slapped the sheet of paper upside down on her tray table.

"I did?"

"Enough conversation for now. Rest."

She rested. For all of five minutes before her mind mutinied on her. She snapped up the paper Brogan left her, prepared to shove it in the tray table drawer. But he'd written something in the margin. It caught her eye.

"My last spring vocal concert. I think you wrote the arrangement yourself. It changed me. Psalm 62:1."

She'd had nineteen spring concerts. Which song? Lucy booted up the laptop and did an Internet search for Psalm 62:1.

Oh.

"Only in God do I find rest; my salvation comes from him."

She could hear the melody dodging in and out of the wrinkles in her brain matter, feel the harmonies making their way to her heart. If she let go, she could see Brogan's choir in front of her. She remembered now. Even the tough football jocks who thought joining the chorus would give them an easy A—and how wrong they were—had tears in their eyes before the piece ended. "Only God is my rock...I won't be shaken, won't be shaken, won't be shaken anymore."

And the student becomes the teacher.

She'd awakened on her side that morning, a shift in her physical world. Her neck less sore. Her throat a smidgen less tight, able to accommodate a sip of water.

Something shifted now, on the soul level. She felt its presence. *Only God.*

Not prognoses. Not successful therapy, either here or with Dr. Hanley. Not the right cocktail of medications. Not an intelligent schedule for her healing. Only God.

And the reminder came from the least likely messenger, if Lucy had no more than the past by which to measure. A future lay ahead. Out there beyond the hospital windows. Beyond the answer to how much of her voice would come back. Beyond the unknown.

Only God.

Compared to the nothingness of the previous day, this day stood in stark contrast. Her afternoon nurse suggested Lucy might want to shower. Yes, yes, yes! She moved slowly, methodically, so as not to jar anything. The nurse wrapped Lucy's arms in clear plastic sleeves to keep the bandages from getting wet. She covered the still-healing stitches on Lucy's chin with a kind of inverted shower cap and secured it with waterproof adhesive tape.

And then the water came. If Lucy's hair had been longer, she might not have tolerated the time it would take to shampoo and rinse her hair. The shorter style—the one Ania had suggested—washed up quickly. Heaven.

By the time Lucy was dried off, dressed, and back in bed, she was exhausted but completely content. Skyping with her family would be less scary for them with her cleaned up.

It had only been a day, but she missed Charlie. She couldn't wait to show him her new trick—swallowing.

The nurse left with a promise to bring her a treat she might appreciate.

Lucy sucked on a raspberry Popsicle with her eyes closed. She couldn't tolerate the ear buds yet, but had her music playing softly enough not to disturb other patients.

"Now, that's a sight."

Her eyes shot open. Dr. Scott!

"I'd say you've taken a few leaps forward in the last twenty-four hours."

If you only knew. She looked for a place to deposit her Popsicle. Eventually he took it from her and tossed it in the wastebasket.

"I promise. We'll get you another when I'm done examining you and talking with you a bit." His enjoyment of the scene couldn't be masked.

"Thank you," she managed.

"Smiling, too? This is a record-breaking day." He peeked at the skin on her arms under the Teflon gauze, at her shredded knees. "Looking good. A little less tender?"

Lucy nodded.

He handed her a tissue. "You have a little something..." He pointed to her mouth.

She wiped around her lips with the tissue. Raspberry. *Nice, Lucy.*

"Ready for this?" He waved a tongue depressor. "Open wide and say 'aaah.'"

Lucy gave him her best fake glare.

"How about," he said, "open as wide as you can and say 'argh' like a hoarse pirate?"

She had to get the comedian. Until today, he hadn't dared to joke around. Lucy took that as a good sign. She opened her mouth, but the argh part wasn't going to happen with her neck stretched upward. And although she'd lived fifty-five years without gagging on a tongue depressor, with the exception of childhood, no amount of "Stay Calm and Open Wide" was going to prevent that from happening today.

"Uh huh." Dr. Scott patted her arm. "Don't worry about it. Not uncommon in these situations, although I haven't seen many cases like yours. I hope you've bronzed that life vest in honor of the role it played in keeping you alive."

God only is my salvation. "Grateful," she said.

"Hmm." Short. Crisp. "Definite improvement there."

"No volume. Or tone."

"Spoken like a musician. I like that, Lucy. Now, let's talk about your scans."

She waited.

"Number one, you're alive."

"It's a start," she whispered.

"A good start. The initial scans when you were admitted showed significant abnormalities. Significant damage. These recent scans show slight improvement. Not all that surprising that you haven't seen a dramatic recovery. Although," he said, "from what I'm seeing"—he pointed toward the wastebasket where her cherished Popsicle languished—"we could take scans right now and maybe see a little less swelling. You're definitely headed in the right direction."

Good. Let's stop right there. Headed in the right direction.

Dr. Scott's intern hovered in the background. His face, she could read.

"I wouldn't want to give you false hope."

An oxymoron, doctor. *False* and *hope* paired are incompatible. Like jumbo shrimp. Or living death.

"So I'll be up front with you." He stopped to smile. Stopped. To smile. Not good. "I've heard you sing."

What?

"You sang at a colleague's wedding last winter. It was"—he shook his head—"the most beautiful.... Stunningly beautiful, really. Pure art."

She signed, "Thank you."

"I think...you'll be...disappointed if you hold onto hope that your vocal capacity will return to that level of excellence. I know that's not what you wanted to hear. I have to be truthful. The gross hyperextension— Let me describe it to you this way. You've seen Silly Putty stretch when you pull it slowly?"

She nodded.

"And when it gets too thin, or if it's tugged rapidly, as with a sudden trauma..."

"It snaps."

"Now, we don't see any actual breaks. That's good. But, that whole area has been so traumatized, there's nothing surgically we can do to bring it back from.... You know what it's like when the elastic wears out in a favorite pair of stretchy pants."

Sir, I haven't worn stretchy pants since shortly after Sam was born.

"Everything else will heal with time. We'll set you up with a vocal coach, a speech pathologist when you're discharged. Maybe tomorrow, if you do well when we remove the IV and switch to liquid meds for pain."

Everything else.

"Do you have any questions for us?"

The intern moved a step closer, but offered nothing except his sympathetic facial expression.

Rest. The word landed like dew on her forehead. It formed a protective barrier the pronouncement couldn't penetrate. Oh, she heard the prognosis. But she couldn't be shaken by it, like she might have been even eight hours earlier.

"We'll get that Popsicle ordered for you. Strawberry?"

She shook her head. "Rasp-raspberry."

35

A merciful sun wore shades the day Lucy moved into her own home. Clinging to Sam's arm for stability, she walked from the car to the door, trailed by family members carting flowers and baskets of cards and balloons and a massive azalea plant. Charlie kept his distance—for her sake—but had the door open before she got there. Again.

"Where do you want to sit?" Charlie asked, standing on the opposite side of the family room. "We may need to get you a recliner. Do you want to use mine? Do you want to see how that works for you?"

"Bed."

"Okay. Sure. I changed the sheets. Sprayed enough Lysol to kill any of these cold germs lurking. I'll camp out here in the family room until I get rid of this. Lucy, you look so good," he called after her as she made her way down the hall to the bedroom.

Life had gone on without her. She came home not as a matriarch but as a houseguest whom everyone bent over backward to serve. *Is anything still the same, God? Anything?*

Olivia slipped past Lucy into the bedroom. She folded back the covers on Lucy's side. "Want the blinds and curtains closed, Mom?"

"Please."

"Something to drink?"

Lucy's chin quivered. Where did all the emotion come from? Gratitude. "Water."

"Okay. Can I help you into your pajamas?"

Lucy held her arms out. Yoga pants and a tank top. Olivia helped her out of her French terry hoodie.

"Oh, right," Olivia said. "That'll work fine. No need to change. You haven't been digging in the dirt or anything."

Thanks for the reminder. She'd barely noticed her flowers in front when climbing the Everest between the car and the front door. Who else in her family would have taken the time?

Olivia stacked pillows to simulate the incline of Lucy's hospital bed. Lucy lowered herself to the edge. Step one.

Sasha stood in the bedroom doorway. She blew Lucy a kiss.

"Come in," Lucy whispered, then signed.

Sasha shook her head *no*, pointing to the bright eyed, emotionless, but sniffling boy clinging to her leg. She signed, "I love you."

"I love you, too."

Sasha and Evan disappeared then and left Lucy and Olivia alone.

"That was pretty sweet," Olivia said.

Lucy lowered herself onto her side, then rolled to her back. Familiar sheets. Familiar bed. Familiar face of her darling daughter.

"Need a nap," she mouthed. "Stronger soon."

"Sleep as long as you need, mama."

Could I sleep until I wake up healed? Of everything?

On the fringes of family life. That's where she hovered. She heard the activity "out there somewhere"—Evan banging on the piano and being shushed for her sake, Charlie coughing and trying to pretend he wasn't, Sam excusing himself to get back to work, Olivia carrying on one-sided verbal conversations with

Sasha and ducking into Lucy's office to talk in hushed tones with Caden. His name rang clear.

Napping meant lying there in the feigned, curtained dark of mid-afternoon, thinking too much.

Caden Cole. The offspring of Evelyn Schindler, the music assassin. Does not compute. Everything Lucy had heard from Olivia about Caden, everything she'd seen with her own eyes, endeared him to her. And his heartfelt apology for his mother's actions? What was Lucy supposed to do with that?

They hadn't been able to take the conversation—such as it was—to any kind of conclusion. *Olivia, keep dating him at least until we know the whole story, okay?* Oh, good glory! If Caden and Olivia grew more serious, got married, Evelyn Schindler would be Olivia's mother-in-law and the other grandmother to any grandbabies they produced!

Apparently, Lucy thought, removing one of the pillows behind her neck, *I've not completely forgiven the woman.*

How necessary was forgiveness, really? And how could God expect her to forgive the woman who nearly single-handedly wiped out the musical futures, the joy futures, of an entire community of children? And Lucy's passion.

She put the pillow back. Her throat tightened when she lay that flat. Or thought those thoughts.

If the school board hadn't eliminated her job, the accident would have.

How had that not registered until now? Her job was finished, RIF letter or no.

Lucy stared past the ceiling. *You probably thought I'd find it easier to forgive now, didn't You, God? But what about the students? The children?* If the program were still alive, Lucy could have been replaced by any number of instructors looking for a position and an innovative program into which to step. Her father's brainchild.

The children. Children.

How far had Ania inched toward faith by now? What were *her* conversations with God like? Anything like Lucy's?

And when did Charlie move their dresser to the other wall?

And how soon before she could stop taking the med that made her slee—

The house that had seemed so swollen—each hallway an artery clogged with people—now seemed gymnasium big as Lucy made her way from the master bedroom to the kitchen filled with sounds of life.

"Mom, did we wake you?" Olivia slapped an Uno card on the table in front of Evan. She sorted by color. Evan's intent looked like road building. But he was engaged. Sasha had her laptop propped on the island and moved from it to the stove where garlic and onion sweated.

"Dad?"

Evan looked up briefly at Lucy's word. Was he thinking about Charlie, Sam, or the way her voice reminded him of a Disney villain?

"He went to the store to stock up on things we thought you could eat. Drink. Swallow. Whatever."

"New book title," Lucy said. "*Eat. Drink. Swallow.*" Enough talking. Appalled. That's how she felt about her limitations, verbally. She watched Sasha add cream cheese to the sweated aromatics, soundlessly, and repented for whining when others' limitations—and freedoms—were so much greater than her own. Those freedoms. Sasha.

Lucy moved into the work triangle, making a serious effort to stay out of Sasha's way. She peeked into the sauté pan.

Olivia said, "She's making alfredo. Did you ever make alfredo, Mom?"

Lucy shook her head no. She stepped out of the way when Sasha left the pan to check something on the laptop again. A recipe? The screen showed an architectural layout. Sasha used her pinky finger to maneuver a change on the layout, hit SAVE, then returned to the stove.

Without thinking what it would cost, Lucy wrapped an arm around Sasha's shoulders. A simple hug. The tender skin on her forearms voiced their distress with the gauze protection gone.

Temporary, Lucy. Many of these things are temporary. A little better every day. No need to mourn the loss of hugging, too.

The smell of food no longer gagged her. Its appeal had returned. She pointed to a covered pot at the back of the stove. Sasha looked to Olivia.

"She's making homemade broth for you. Bone broth, whatever that means. Supposed to be super good for you. The woman's a genius in the kitchen." Olivia bent to retrieve Uno cards that had showered onto the floor.

Sasha opened the freezer door and pulled out a small, wildly colored plastic tray. "Oh, yeah," Olivia said. "She made you homemade freezer-sicles from the contents of a variety pack of Evan's juice boxes. Grape, cherry, apple..."

Lucy took the extended tray and chose one with a cartoony airplane handle. She signed, "Thank you."

Evan noticed the exchange and bounded over, grunting his demand with his hand opening and closing, gimmee-gimmee style.

"For Grandma," Sasha signed.

Lucy touched Sasha's arm. "May I?"

Sasha nodded. Lucy turned to Evan, who looked at the inanimate object rather than her. "Ask nicely, please." He fussed a moment, then said, "Please one of those." She held the tray out to Evan, who chose the same flavor she had.

Olivia rose from the table and grabbed a wad of paper towels. "You two better eat those things outside." She held the door open. First Evan then Lucy stepped out onto the deck.

Lucy stopped abruptly. Goosebumps danced on her skin.

"Ta da!" Olivia said. "How do you like it?"

A deck-top screen house.

"Dad and Sam put it up. Sam bought it for you, us," she said. "Just finished late last night. Isn't it cool?" She unzipped the door to let the Popsicle bearers in, made sure both were settled in patio chairs, then zipped herself out.

Lucy and Evan sat focused on their rapidly melting iced treats, sure to spoil Evan's supper but probably the bulk of Lucy's meal. She couldn't read where Evan's thoughts roamed. Hers entertained a picture of her son and Charlie working into the night to erect the screen house. Shade. No bugs. A serene spot.

She swallowed past the lump in her throat.

Lucy finished her treat first and watched as Evan divided his attention between the dripping ice and the rhythm his fingers tapped on the patio table. It didn't take long for Lucy to distinguish "Twinkle, Twinkle, Little Star." One finger. Two taps, two taps, two taps, one. Rest. Two, two, two, one. Rest.

The premise held true, as it always had. Music busts through barriers. Music matters. It matters. Maybe she'd write about that.

Someone, probably Olivia, had put the azalea in the center of the patio table. Tomorrow, weather permitting, she'd sit out here and write thank-you notes. And reread the messages from former students who found innumerable ways to reinforce how music had mattered in their lives.

It was time to set old expectations aside. But not without gratitude.

Lucy drummed an accompaniment on her side of the patio table. Evan didn't look up, didn't acknowledge her, but smiled.

Olivia stuck her head out of the house. "Mom? Somebody named Marta called. Is she from your book club?"

Lucy nodded.

"She wants to come visit. Says she has something for you."

Was she ready for company? Tomorrow?

"I asked if her gift would spoil. She said, get this, 'Oh, no, dear. Quite the opposite.' I wonder what she meant by that."

Despite the after-effects of the grape icicle, she warmed inside.

"I asked her if she would mind waiting until early next week, then. Told her you were doing fine but needed to build up your strength. She said, 'Don't we all?' But then she said that would work out well. So she, and I guess a couple of the others from the book club, are going to stop in sometime on Monday. Unless that's not okay with you. I took her number, in case you didn't have it."

Lucy made the international sign for *okay*.

Sitting there in the sun- and friendship-warmed day, the music-warmed, family-warmed, peace-hemmed day, Lucy plotted what she'd serve the HHATT club when they came. Tea and...?

And how she'd introduce her husband. "This is my husband, Charlie, who sneaks messages into potted plants and lays down his life for me, volunteers to abdicate his recliner, sleeps on the couch, and installs a screen house by flashlight because he loves me."

So many words. She'd have to write them down.

Charlie was a blessed man. Three women in the house, and only one of them could talk. What would they do when Olivia moved out? Could she and Charlie keep them all forever? No. How would they find their own place in the world? She prayed again that where they landed wouldn't be far from Cottonwood Street.

Lucy watched Evan, hawk-like, in a loving sort of way. She assumed the potted azalea was too heavy for him, but didn't want to give him a chance to find out what happens when something so beautiful crashes to the ground.

Sometimes the petals fall. Sometimes the roots get damaged. Sometimes you have to replace the container. Sometimes you make compost and start over.

36

Charlie arrived home with groceries and a hazmat suit.

Unofficial hazmat. A painter's white Tyvek hooded jumpsuit—Ed Asner shaped, a medical facemask, and purple surgical gloves. "I couldn't stand not holding you."

Charlie, Charlie, Charlie.

Lucy could have kissed him, but she didn't dare.

Olivia said, "How about if I put the groceries away and you two...uh...spend a moment on the couch. Sitting. Sitting upright on the couch. Together. Sasha will have supper ready soon. Very soon, so..."

They sat side by side on the couch, his protective and protected arm around her shoulders. He offered his gloved hand. She took it.

"Life got a little crazy, didn't it?" he said, his voice low and soothing.

She nodded.

"I'm glad you're home."

She swallowed and then rasped, "I'm glad we're us." She leaned against his arm, not remembering if anything still hurt.

In the hospital, Charlie had read her one of his, "Hey, listen to this" trivia bits from a magazine he'd borrowed from the visitor's lounge. "Did you know," he'd said, "that the weight of a burden you hold at arm's length doesn't matter as much as how long you hold it?"

"And that matters, how?" she'd written on the now absent whiteboard.

"If you hold a heavy object for a few seconds, no problem. Ten minutes? Muscle strain. An hour? Major problems. I'm paraphrasing," he said.

She hadn't had the strength to ask the point. He offered it anyway.

"Even for burdens we're forced to carry, we have to set them down once in a while. Frequently. It's the only way we'll have the endurance to pick them up again and keep going longer."

She thought she'd read that somewhere else, using a glass of water as the illustration. It never resonated as it did now, hearing it through Charlie. They had a burden to carry. More than one. For a moment they leaned back on the couch together and set it down.

Sam had whispered a verse to her on his first hospital visit. She'd been foggy-headed, but she remembered it now. Word for word the way he'd said it. Jesus said, "Come to me, all you who are struggling hard and carrying heavy loads, and I will give you rest." At the time, she wondered if he'd had a premonition that she wouldn't survive this, that the home to which she'd head was spelled with a capital H. Rest. So many meanings.

Lucy matched her breathing to Charlie's, except for his occasional barely suppressed cough. He looked like an investigator for the CDC. This time the disease was behind the mask, not in front of it. The aquarium burbled. Dusk deepened outside the wide window. So much was wrong. But all was well.

Divested of his hazmat suit, Charlie and Lucy ate their supper in the screen house. He figured—and she agreed—that the fresh air carried away anything faintly resembling contagion. The chicken alfredo Florentine looked and smelled delicious. Lucy contented herself with a half cup of the bone broth and a glass of water Sasha had infused with strawberries and cucumbers, then strained. Lucy appreciated the hint of flavor without the texture.

Every swallow seemed a little easier than the last. She still felt like a woman with a linebacker's neck, inside and out. But she was upright. Breathing. With the people she loved.

Tomorrow? Applesauce.

Olivia had set their patio table with a votive candle in a turquoise canning jar because, she said, it might get dark out there. A small votive. Small gesture. Large impact. It turned a meal of broth—no bread—into a soul-refreshing experience.

"Aren't you going to get hungry before bedtime?" Charlie asked, sopping up the remainder of the Alfredo sauce with the last bite of crusty garlic bread.

"Small meals," she said. "More often."

"Smart."

"Plus early bedtime."

Charlie wiped his mouth with the cloth napkin and set it as far away from her as he could reach. "I hear you. Haven't slept well since..." His voice and gaze trailed off.

How different conversations seemed when she listened more than she spoke, of necessity. How long could she maintain that grace if her voice got stronger? *God, I want to be healed, but I don't want to be the same as I once was.*

"One misstep," she said, fully conscious of sounding like her husband's much older chain-smoking brother than his wife.

"That's how most disasters start," he said, still facing the gathering darkness. "I should have—"

She slapped the table. Not hard, but enough to draw his attention. "No regrets. Either of us. We can't afford them."

He coughed into his elbow, then leaned back. "It's possible I need hearing aids. Especially if this is as loud as you can talk from now on. I thought you said, 'No egrets. Either of us.'"

"Could be your cold."

"Could be I'm old?" he mimicked, rasp and all, then winked. He reached to touch her hand, then thought better of it.

They sat that way, in the silence and the honey-like sweetness of nothing to say and no need to say it, until the back door thunked open. "Humidity," Olivia said. "No one believes the kind of humidity we have here in Wisconsin in August." She set Lucy's cell phone and a portable Rubik's cube-sized speaker on the patio table near the azalea. "Mood music." Without another word, she removed their plates and slipped back into the kitchen.

"That girl..." Charlie said.

"We did a few things right, Charlie."

A woodwind quintet added its baroque notes to the outdoor scene. Lucy reached for her phone to choose something else from her playlist, for Charlie's sake. Not his style. He waved her off. "No. That's kind of nice. Leave it."

Kind of nice. She should have had Charlie come into her classroom to give a testimonial about the nuances of baroque.

"I had," he said, "an interesting senior moment today."

She tilted her head. Listening mode.

"Well, I saw this guy in a truck across the parking lot at the Pick 'n' Save. Mike Middleton. Remember him? Haven't seen him for years. Still had that same faded hat he used to wear when we had our mill department fishing tournaments. Same beard down to his belt."

"Charlie."

"Not quite. You remember?"

She did.

"So I jogged across the parking lot to say hi."

"You jogged?"

"That's what I called it. Anyway, he rolled down his window and I said, 'Hey, buddy! Great to see you again!' And he says, 'You, too.' And that's when I realized it wasn't Mike. Nope. Not Mike. But he says to me, 'How's life?' and I thought it would take too long to tell him so I said, 'Ups and downs.'"

Eloquent. And disturbingly true.

"But then he says, 'You still doing all that painting?' Lucy, he didn't have a clue who I was either! Can you picture this? So I said, 'Painting? Every chance I get.'"

Lucy doubled over with laughter, barking like an act at SeaWorld.

"Careful now," he said. "I didn't mean to give you a spasm."

She could hardly catch her breath, but didn't mind. Bach, Scarletti, Vivaldi, and her Charlie. On this night, it made sense. It made a life she could live with.

The process of writing thank-you notes exhausted Lucy. The caring behind each gesture from a community she'd thought uncaring or at best misguided—since, bulldozer that she was, Evelyn Schindler hadn't acted alone—overwhelmed her.

Charlie had read Lucy all the messages in the cards when she was in the hospital. At the time, she'd thought his occasional tears were because of how broken she was. And that may have been part of it. But as she revisited the sentiment of those same cards and notes now, she could feel what he felt. Complete awe that she and music—music and she—*music* she had the privilege of introducing—had touched so many lives.

Maybe through Evan, that privilege of introduction wasn't over.

Wasn't it Nietzsche who said, "When we are tired, we are attacked by ideas we conquered long ago"? Or was that Dr. Hanley? A cloud she'd thought conquered hung itself above her post-concussion head. She'd found her song again, but now couldn't sing.

From her spot of serenity in the screen house, she watched a vapor kind of cloud scud across the sun then continue on its path elsewhere in the sky. The cloud, real and visible, quickly lost its power to darken the scene.

Her personal cloud, true and legitimate, didn't respond immediately when she told it to be on its way. It didn't move until she read the next note. From Ellie and her mother. The mom had enrolled Ellie in private lessons through an internship program at the university. The note included an invitation to Ellie's Christmas recital in December. What faith! Lucy keyed it into her online calendar. Lucy recognized Ellie's unique penmanship—full of curlicues and life—on the handcrafted fabric bookmark tucked into the note:

> *After you have suffered for a little while, the God of all grace, the one who called you...will himself restore, empower, strengthen, and establish you.* —1 Peter 5:10

Of all people, the thirteen-year-old with scarred lungs offered the teacher with a scarred throat and heart a lungful of hope.

A student—a junior at the high school in Woodbridge when Lucy left Willowcrest School in May, a young man who would be a senior in a few weeks, an exceptional musician who anchored the tenor section of choir and the baritone section of her band—sent what looked like a typical get-well card. He let the card do most of the talking. But beneath his barely readable signature, he'd printed, "Mrs. Tuttle, don't forget to *play the rests.*" He'd underlined the last three words.

Play the rests. The rests.

The temporary rests between notes. The forced rest of a hospital stay. The gut-wrenching "rest" dictated by the RIF letter. The vocal rest. The temporary—she hoped—rest from pursuing her passion. The rest—remainder—of her life. *Play it. Don't let up the intensity, the focus, the concentration, the emotional connection.*

She'd succeeded in convincing at least some of her students over the years. So many mentioned that phrase, or one of the other *Tuttleisms*, as they called them. Lucy laid her hand on the stack of notes and cards and let the tears flow. Again.

"Mom," Olivia said. "I'm leaving for work. Are you okay?"

"I will be."

"Do you want me to call in sick? Get a replacement?"

"No, hon. I'm fine." Lucy stuffed the cards and her writing materials into the tote bag she'd once carried to school every day.

"Are these done? I can mail them on my way to work."

"Would you? Thanks."

Olivia stacked the completed envelopes. "Stamped and everything? You've been busy."

"I've been blessed."

"Your voice sounds a little more gravelly than usual today. Are you sure you're not overdoing it?"

Lucy pointed to her throat. "Emotion."

"Well, don't overdose on that either." Olivia kissed her mom on the cheek and said, "Working until nine tonight. Dad's in there watching fish with Evan. Text him if you need him."

Text her husband, thirty feet away. Her nod promised she would.

Her few tomato plants were starting to produce. She saw flashes of red among the wrinkled green leaves. Lucy left her screen house sanctuary and walked the stone path to the vegetable patch. She didn't know how Sasha found the time, or maybe it was Charlie, but the garden looked in better shape than she expected for this late in the season. A twinge of jealousy over climates with longer growing seasons dissipated when she bent to cup her hand around a ripe-enough-to-pick tomato. Two tomatoes? No. Double vision. One. Two again. She stood slowly and let her equilibrium have a chance to orient.

She'd ask Charlie to come pick the tomato later. Oh. It was in her hand.

How long since she'd eaten? Something to eat, then she'd lie down for a while.

Lucy made her way slowly back to the deck, up the couple of steps, and to the house. The air conditioned interior helped. More clear headed. Less like she was going to pass out.

She left the tomato on the island and let her bed and pillows lure her toward her nap destination.

"Are you okay, Lucy?" Charlie asked as she maneuvered through the family room.

Everybody kept asking that. "I will be."

37

Monday morning brought driving rain. She could hear it, and see it through the space where the bedroom curtains didn't quite come together as she lay listening to the day awaken, to her soul awaken. The garden and flowerbeds needed the rain. Charlie didn't mind mowing less frequently, but he did take pride in a well-manicured lawn. Rain. How long had it been? Refreshing, despite its intensity.

Rain brought an accompanying dilemma. Lucy scooted closer and rubbed Charlie's back. If the HHATT club couldn't meet on the deck, where in the house would they find any privacy, any freedom to talk freely? She caught the irony. She would, for once, be talking little. But still...

"What time is your book club coming?" Charlie said as he rolled toward her.

Lucy held up ten fingers.

"Oh, LucyMyLight."

"What?"

"I won't be here to help you serve them. Evan has an appointment with his speech pathologist—I know, right?—so Sasha asked if I'd drive them. It'll take us a couple of hours at least. I thought I'd get them some lunch on the way home."

"Thank you for doing that."

"My pleasure. Just glad I'm available. You'll be okay without us here?"

She counted on her fingers, "Carole, Marta, Angeline..."

"Oh, sure. They'll know not to expect too much from you."

Lucy glared at him.

"You know what I mean. Because you're recuperating. They'll be sensitive."

More sensitive than some men I know.

"I'll get the cookies all laid out on a plate and get the tea things ready and on the table in the kitchen before we go."

"You made cookies?"

"Uh, yeah. They look exactly like Oreos."

"Thank you."

"I'll put them on that plate you have that's shaped like a book."

"What?"

"*Book* club?"

Lucy smiled. If he only knew.

"And I'll set out the tea cups with the musical notes on them because, well, we have more matching ones of those than anything."

She played with the silver around his ears. Maybe he'd trust her with scissors, shears, soon.

"Unless you want the one Martin got me with the big-mouth bass."

She could see the table set with that monstrosity.

"Nah," he said. "Not enough of them to go around."

"Going to get up now."

"Me too." But he lingered. So did she.

"Did I snore last night?" he asked.

She nodded.

"One good thing. My snoring won't ever disturb Sasha."

Oh, I'm pretty sure she can feel the vibrations clear down in the basement.

"Did Sam tell you they're looking at another apartment this weekend?"

"Sasha told me. It's farther away than I'd hoped they'd be."

Charlie remained quiet for the space of two breaths. Two morning breaths. "Part of me hopes it won't work out," Charlie said.

"I know what you mean." The quiet she'd longed for was far from that, but so...God-hemmed.

Neither said anything more.

The sound of the rain. The sound of Charlie's steady breathing that could morph back into snoring if they didn't decide to get up pretty soon. The sounds of a young family stirring in the lower level. Sounds of home. Music to her ears. Embraceable music.

"Race you to the shower," she said.

His eyes popped open. "You don't race."

She crawled out of bed. "Won't that make it all the more embarrassing," she rasped, "when I beat you?"

"So, it's like the game Twenty Questions?" Angeline asked.

Lucy set the carafe of hot water on the table and pointed to the small basket of teabag options. "Kind of," she answered. "Don't have much oomph behind my words yet."

"And doesn't it hurt when you talk?" Marta asked. "I would think it would hurt."

Lucy nodded. "Ask questions I can answer yes or no. It helps."

Carole clasped her hands in front of her on the table, her face pouting. "But we all want to hear the story."

Girls just want to have fun, Lucy thought. She mimed—Charades style—two people in a canoe, paddle, paddle, paddle. Stop paddling. First person. She mimed stepping out of the canoe, slipping on the mossy-covered boat ramp, slapped her hands together to mimic the impact and bent the fingers of one hand back to help them picture what had happened to her body.

She showed them the scar on her chin, the scraped flesh healing well on her forearms, and pulled up the hem of her capris to reveal her healing knees. She indicated a dizzy, throbbing, aching head. She used her hands to form a giant neck around hers and grimaced. With her as-yet empty tea cup, she sipped imaginary water and grimaced again, shaking her head. She stiffened, as if lying in a hospital bed, and counted with her fingers the number of days.

By that point in the story, Marta could hardly contain herself with laughter and Carole excused herself to use the restroom. When Carole returned, Lucy said, "The end," and curtsied.

Angeline chided the others. "This is no laughing matter. It was a horrible thing."

"Parts of it," Lucy said.

"Like," Angeline sniffed, "the way it left her voice sounding like that. No offense."

Lucy laughed. "None taken." She poured hot water over the jasmine tea bag in her cup.

Carole said, "It's not really the end, though. There's more to come yet. You're still healing. Right?"

"Right. Don't know how much stronger my voice will get."

To a person, each woman grabbed a tea tag and dunked it in the steaming water. Noiselessly.

"I learned a few things, though." Lucy said.

"How to talk like a pirate?"

"Marta!" Angeline slapped the grinning woman's arm.

Lucy looked into the faces of these three very different, very precious women. Where would she be without them? Dr. Hanley had known just what she needed to not let her marriage get lost in the fumes of her job loss.

"How can we help you, Lucy?" Marta again. Serious and sympathetic.

"Keep doing this," she said, indicating the gathering around her table. "And this." She held up her pink patent leather journal. Lucy opened hers and the others followed. "I have something."

"Was this written," Carole asked, her face a mash-up of concern and playfulness, "while you were still on the high-powered drugs, dear?"

Angeline crossed her arms over her chest and shook her head from side to side.

Lucy took a sip of tea and read, "Don't be afraid to be honest with your husband." She paused. "It's taken me a long time to figure that one out. Still working on it."

"Oh, that's good," Angeline said. They all wrote.

Lucy took a sip. The others grabbed an Oreo in one hand, their pens in the other. She swallowed her tea, thanking God she could, and read, "Don't be afraid to be exceptionally kind to your husband."

The raised eyebrows around the table didn't deter her. "Don't be afraid to be away from your husband when you need the focus only solitude can bring." Her throat ached, so she paused longer than it took them to copy the sentence in their journals.

"Is that it?" Carole drew a line in her book.

Lucy shook her head no. "Don't be afraid to be different from your husband."

Nods around the table.

"And," Lucy said, "don't be resistant to counting on those differences, rather than resenting them." Another sip of tea. "Or laughing together about them."

"Which traces back," Marta added, "to our humor homework week. Our house hasn't been the same since we reintroduced the humor part of our relationship."

"What did I miss?" Lucy asked. "When you met and I wasn't there?"

"A whole lot of fun," Marta said.

"Dr. Hanley suggested," Angeline interjected, "that we start working on our profiles for 'The Wife I Want to Be.'"

Marta chuckled. "Like I said. A whole lot of fun."

Was Lucy the wife she wanted to be? No. Closer, but no. "What kinds of things do we write down?"

Angeline took a deep breath. "I want to be the kind of wife who presumes the best, who doesn't jump to conclusions about my husband's motives. I've been wrong about that too many times before."

Lucy thought of a few while the others gave their input. Prays more diligently for her husband. Gives more instant passes. Is more cheerful in the morning and not centered on getting breakfast made, but on starting her husband's day and their marriage day off right. Cares more about relationship repair than about proving how right she is. That one made her physically flinch.

In one of their sessions, Dr. Hanley said, "A woman's whole life can change the day she starts telling the truth about what makes her happy and quits making people guess. That's not selfish. That's communication."

Lucy jotted down the suggestions others made. She'd study them later to see if they belonged in her own description of The Wife I Want to Be. The Woman I Want to Be. Ooh, she raced ahead to future lists of The Mom I Want to Be. The Grandma... The Friend...

"Personally," Carole said, "I want to be the kind of wife who wakes up grumpy... and tells him to sleep a little longer."

The table rocked with laughter.

"Before this gets too academic, ladies," Carole said, "let's give Lucy our get-well present."

Lucy shook her head. As if that would stop them.

"Presents," Angeline noted, beaming as she emphasized the plural. "Here's mine." She handed Lucy a tissue-paper-wrapped package. An artsy coffee table book called In the Silence. Stunning photography with one word of text per image. Peace. Tranquility. Solitude. Song... Song, in the silence. Lucy understood the concept better than she would have a few months earlier.

Carole's gift—a sterling silver bracelet inscribed, "Your life's song never ends."

In true Marta fashion, her gift to Lucy needed explanation. Lucy ran her fingers along the edges of the beautifully engraved gilt frame. "Hand-carved?" Lucy asked.

"I don't know. It's more than a hundred and fifty years old, so it's possible."

"Oh, Marta."

"I didn't pay much for it, if you're worried about that. It came from my great-aunt's attic."

"Nice one, Marta," Carole whispered.

"It's what it frames that counts," Marta said. "See?"

A thick black bar attached to a single thin black line.

"This way," Marta said, taking it from Lucy, "it's a half rest. This direction"—she flipped it end for end—"it's a whole rest."

So it was.

"When you die—"

"Marta!" Carole and Angeline voiced their disapproval in unison.

"Later, much later," she added, "your family can flip the painting to show how you finally got your full rest."

"That's disturbing," Angeline said.

"No, it's meaningful." Lucy reclaimed the framed painting. "Only a fellow musician would think of that. Thank you." She dabbed at the corners of her eyes, marveling that tears no longer made her head hurt.

"We'd better get to the group gift before this turns into a sob fest," Carole said. She pulled an envelope from her purse. "There you go."

"It's a gift certificate," Angeline offered.

"Let her open it."

"It's for ESL classes," Carole said when Lucy didn't respond.

"Well, that would be interesting, but I don't think she needs English as a Second Language instruction. She does pretty well with English." Marta shook her head.

"ASL," Lucy said.

"Yes."

"I need this." Who would have known how much she longed to communicate more fluently with her daughter-in-law? She'd planned to start ASL lessons. Then life got even more complicated than it had been.

"Charlie went in with us," Carole said. "We were going to bring you flowers, but—" She indicated the veritable greenhouse full of them still fresh enough, most of them, from the hospital.

"Charlie helped with this?"

Angeline crossed her hands over her heart. "He said you could take the classes, then he'd learn from you. Look at the description of the last set of lessons in the brochure they included."

Lucy scanned the list—conversational, business, emergency uses...practical application. She knew it would take years to become truly fluent. But this crash course would—Oh. Worship. Lessons specifically addressing signing worship.

Someone like Lucy, who could hear the words, hear the music, the rhythm, the emotion.... Someone like Lucy, who connected more deeply every day to the One the sermon talked about, the One the songs praised.

She fingered the edges of the gift certificate with her name on it. Signed by all three, plus her Charlie.

"I forgot something in the list I shared earlier," Lucy said, pointing to their journals. "Start with a good man."

38

Evelyn Schindler was dead and it was not Lucy's fault.

Lucy didn't even know the woman had Acute Myelocytic Leukemia. Caden had only heard about her struggle a week before she died. Evelyn hadn't known what was wrong until the middle of July, people said.

The news stunned Willowcrest. It flattened Lucy. For all the devastation Evelyn caused, she was—hard as it was to believe—Caden Cole's mother. In the few times Caden had come to the house with Olivia, Lucy and Charlie had seen a caring, compassionate, intelligent man fully immersed in his love of music. He couldn't imagine, he'd said, a world without music education. Raised by Evelyn Schindler?

"More music education, not less," he'd said.

Lucy didn't want to turn Olivia's dates into musical politics or badmouth his mother. She was beyond that. She'd been determined to build a bridge of forgiveness from her side of the chasm for months—some days more successfully than others—and intended to bridge the entire gap for her side if necessary. Evelyn's death meant she'd never hear the hardest words Lucy had been practicing.

Caden canceled his Toronto tour to stay in Willowcrest long enough for funeral preparations and all the attendant responsibilities when the last surviving parent dies. Olivia was by his side for much of it, at his request. She thrived in that difficult role, as Lucy and Charlie knew she would.

The new school year was in full swing. Lucy's healing had brought her to a good place. She'd probably spend most of the fall getting back up to normal energy level, but her voice was so much improved that she no longer fielded pirate jokes from her family or friends. She sounded like a woman recuperating from laryngitis.

She couldn't sing one note. Not one. Her throat collapsed with every attempt. She relied on the instruments she played, especially her piano, to communicate what she'd always wanted to say through music. She and her baby grand got to know each other much more intimately after the accident.

Until his mom's death, Caden's date nights with Olivia often included his guitar, her piano, and a small electronic piano—complete with headphones—they'd found for Evan. He made music day or night when the mood struck him, which was often, according to Sasha. Charlie sat in his recliner, feet up, sometimes humming along, sometimes content to be captivated.

Now, date nights were centered around funeral preparation.

"I want to write a song in honor of my mom, Mrs. Tuttle."

"You're a songwriter and performer," Lucy said. "I think that's a great idea. No one will think it's out of place for her funeral." A few of the most vocal against her decision might, but they weren't likely to attend Evelyn Schindler's funeral.

"Would you help me?"

Charlie righted his recliner at that. "Whoa. Hang on."

"Charlie."

"Son, you're a fine young man. You've been good to Olivia and I know she means a lot to you."

"Charlie!" How threatening could Lucy sound with only half a voice?

"Mr. Tuttle, have I done something wrong?" Caden's expressive, emotive face did what it did best—emote.

"Not yet. But I don't want to see it happen. Do you know what it would cost this woman to help you write a song to honor your mother after the way she was treated?"

"Charlie, stop. Please."

Caden backed up a step. Olivia had gone to her room to change, so was likely only hearing snippets... and likely changing at lightning speed.

Charlie rubbed the tops of his knees. And stared at the floor. "I apologize for sounding abrupt there, Caden. I love this woman. And I've watched her go through misery over your mother—"

"—and the rest of the school board, who thought they had no other options," Lucy added.

"Well, they were wrong. How they ripped the music right out of the hearts of those kids. And her." He pointed Lucy's way.

Olivia entered the room just as Caden sat cross-legged on the floor in front of Charlie's chair. "Mr. Tuttle, I don't know what to say. But I'm sorry I offended you both. I can't explain all that motivated my mother to be the way she was. But I think I understand part of it."

"I thought you were on the side of music." Charlie leaned back and started the recliner rocking.

"Music doesn't have a side," Lucy said. "That's part of the wonder of it."

"Exactly, Mrs. Tuttle."

Olivia sank to the floor beside Caden.

Lucy was tempted to say, "Let's all go get ice cream." But the boy had just lost his mother, saint or not. And Lucy's husband was showing the first signs that he might understand what cutting the music program meant after all, the difference between a job and a passion. And her daughter looked completely crestfallen. No. Ice cream wasn't going to help anything.

Charlie stopped rocking. "Is it wrong for a man to be protective of his wife?"

"When she *needs* protection, no," Lucy said. "We haven't even heard his story, Charlie."

Charlie mumbled something under his breath.

"Charlie!"

"I'm praying. Give me a minute." He opened one eye. "Okay. So I may have overreacted. Let's hear the story."

Not surprisingly, Lucy's throat tightened, as it did in any conflict. She implemented the relaxation techniques Brogan taught her, much easier to follow than the ones her new speech and language therapist recommended. Maybe Brogan's musical background made the difference.

Caden took Olivia's hand. "I believe my mother had an unconscious aversion to the arts because she believed music stole me from her."

You could hear the *shadow* of a pin drop before the metal hit the floor in that stark stillness.

Charlie leaned toward him. In a voice nearly as soft as Lucy's but much richer, he asked, "Was she right?"

"It didn't have to be that way. I was all she had. That's how she saw it. So when I felt the pull to a career in music, and that meant leaving the small town where we lived at the time, a place a lot like Willowcrest, she felt I'd abandoned her, that I'd chosen music above her. That wasn't true at all. But she gave me an ultimatum. Give up my passion and come home, or don't come home."

Charlie shot a glance Lucy's way. She saw him through her fingers pressed against her face.

"I was nineteen. She was lonely and bitter. Maybe if I'd been more mature.... No. I don't know how I could have made any other choice." He bounced his hand, with Olivia's tucked inside it, on his blue jeaned knee.

"I think after a while," he said, "she realized she was the one keeping us apart. But she was so stubborn, she wouldn't admit it. She could be stubborn."

"Oh, son, we do know that."

Lucy watched Charlie's posture for clues to what he might say next.

"About this song..."

"A song *in* honor of my mother, not specifically *to* honor her, although I'm sure that sounds like splitting hairs."

Lucy sat on the arm of Charlie's recliner. "I don't know what input I could have."

"Help me get the words right. I only have two more days. The last thing I want to do is *dis*honor her."

Olivia put her arm across his shoulders. "Let them see what you have so far."

"I didn't bring my guitar in the house."

"I'll go get it," Olivia said. "Mom, we may need ice cream after this."

Charlie leaned to pat Caden on the shoulder. "Do you have a title?"

Caden pulled a guitar pick from his shirt pocket. "What She'd Tell You Now."

"Zip me up, Charlie?"

"I still can't believe you're going to this funeral." He tugged the dress's zipper into place and fumbled with the hook at the top.

"I can't believe you're going with me." She turned and straightened his tie.

"Body guard," he said.

She put her hands on her hips. "Oh really?"

"Actually, I'm going for Caden. *And* you."

Days like this when Charlie wore cologne...and a tie...and came out of himself to be there for someone else...

"Hey, what's that hug for? You'll muss my wrinkle-free shirt."

"I appreciate what you're doing. That's all."

"Nice to hear that." He held her at arm's length. "Now, I know enough not to say this sort of thing ordinarily, but are you sure that's what you want to wear?"

She looked down the length of her mid-calf dress. "What's wrong with it?"

"You still haven't gotten back up to your pre-broth weight. It looks kind of big, baggy."

"Baggy. Swell. I'll just have to be all sad-looking then."

"You have other dresses." He nuzzled his face into the curve of her neck.

"It has to be this black one."

Charlie huffed. "I don't think they stick to that funeral-black rule around here anymore, do they? Except maybe for immediate family. No, I don't think even then. The last time I saw an all-black funeral was when Dr. Phil had a—"

"Charlie? Would you be happy if I added a belt?"

"Might help. I'm only thinking of you, LucyMyLight." His eyes twinkled. He really was thinking of her.

"There. Is that better?" She preened for him.

"A black belt?"

"Has to be. Not karate style black belt, but.... You'll see." She adjusted it so the sequins on the clasp sat in the small of her back rather than in front.

"You going to wear any jewelry with that?"

"Not today."

He tucked a clean handkerchief into his suit coat pocket and said, "There's your fashion statement, right there."

Sam and Sasha couldn't risk losing the apartment they were going to view for the second time. Olivia rode to the church with Caden. So the house was eerily empty when Charlie and Lucy exited and locked the door.

"Temps are cooling down a little every day," Charlie said.

"Always does this time of year." She breathed deep of the faint scent of sunburned leaves. Cottonwoods.

"How much do you miss it, Lucy?"

"School?" She deadheaded a mum on their way past the flower-bed. "I would have said excruciatingly so. But I'm... more at peace about it for myself. I still ache for the disadvantaged students."

"Just the disadvantaged ones?"

She stopped at the passenger door and waited while he reached to open it for her. "They're all disadvantaged if they're not introduced to music."

"I read that somewhere the other day."

"Read what?" She buckled her seatbelt, her mind on Caden's part of the service.

"About music keeping students' minds agile and efficient."

"Are you preaching to me, Charlie? Because that's not really necessary, you know."

He started the engine and backed out of the driveway. "It's fascinating, really."

"I know." She smiled and leaned against his shoulder.

"Did you know that students involved in quality music programs have lower dropout rates, higher standardized test scores, better English scores, they're better in math, and, get this," he said, "are better problem solvers?"

Lucy pressed her hands to her heart. She caught a whiff of his cologne lingering on her hands. "Yes, Charlie. I did." *Be still my heart.*

"How can any community afford to drop its music program with outcomes like that?"

"Are you done?"

"I have more." He wiggled his silver eyebrows.

"Can you save it for some other time? No matter how we felt about what she did, Evelyn Schindler deserves a respectful funeral. And Caden deserves our undistracted support."

"And you," he said, slowing for the first four-way stop, "deserve to do what you were called to do."

"I will, Charlie. I will."

39

It came as no surprise that the day of Evelyn Schindler's funeral hadn't dawned. Not that anyone could notice. The fog and mist had made night ooze into a slightly lighter version.

Lucy and Charlie held umbrellas over their heads on their way to the Traverse. It seemed pointless, since the dampness seeped as much from underneath as overhead. But at least they had their umbrellas for later if the damp mist turned into legitimate rain.

Evelyn's church lay a half-mile out of town, along the river, which made it especially treacherous for attendees unfamiliar with the twists and turns of the river-hugging road on a day when visibility was so poor. Charlie and Lucy took it slow, but still braked sharply when tiny red specks turned into taillights of a vehicle they didn't know was ahead of them.

The stained glass windows would have been beautiful on any other day. The fog outside dulled them. Though Charlie and Lucy didn't worship here, they'd attended special events, concerts, and a couple of weddings, so they were familiar enough with the layout. Lucy and Charlie both stopped at the restrooms before making their way to the front of the church.

No casket. Evelyn specified that and many other details in her will. But the front of the church looked clogged with flowers and plants. Upstanding member of the community and all.

Every time a less-than-gracious thought crossed Lucy's mind, she focused her attention on Caden, whom she'd come to love like a son. Caden grieved the loss of his mother and of what might have been if they'd reconciled sooner than those last few days before her death. Lucy wondered if he would have matured to the man he was if he hadn't chosen a life separate from his mother's influence. He'd pulled free to follow what he believed was a calling on his life. Lucy, for one, was proud of the man he'd become. The man and the musician.

He'd asked her and Charlie to sit near the front. As uncomfortable as she might have thought that would make her feel, it seemed appropriate. Right. They were there to support him, and to show God's power to heal. Physically, emotionally, and spiritually, she stood as evidence.

Evelyn's pastor read the eulogy. Long list of accomplishments, some of which surprised Lucy. She hadn't seen the philanthropic side of the school board president. And she hadn't known until the pastor mentioned it that Evelyn had a collection of Caden's songs, appearance flyers, CDs. She'd followed him silently and in her final hours told those around her bed that his music brought her great joy. She'd never told him.

The Willowcrest teachers—so many familiar faces, dear friends—sat in a huddle. Some waved at Lucy when she glanced back at their rows. She looked forward to catching up with them at the meal following the service. A couple of months ago, that idea would have been too painful. She'd come so far.

The pastor read Scripture—familiar passages about "the tent that we live in on earth" and "whoever believes in Me will live, even though they die." He spoke for a few moments about eternity and God's love, and then invited Caden to the platform.

Lucy rose when he did. Charlie must have thought at first that she needed another bathroom break. But she followed Caden to the platform and stood beside and a little behind him, her hands clasped in front of her until he began to play and sing.

She signed as he sang, her hands and healed forearms visible against the contrasting black dress, she hoped. Lucy faltered once when she caught a glimpse of Sam, Sasha, and Evan slipping into the back row, but pressed on, concentrating on getting the motions right, and making them flow with Caden's song. Making her hands sing.

Charlie had tears coursing down his cheeks. She didn't dare watch him. She faced out and up, beyond the balcony, offering her silent song to the God who had seen her through, who rescued, healed, sustained. The God Caden sang about.

The verses were his own, a sensitive message from son to mother, then a message she might have shared if she were there. The chorus was an adaptation of a quote from Kahlil Gibran:

> *Only when you drink from the river of silence shall you indeed sing. And when you have reached the mountaintop, then shall you begin to climb. And when the earth shall claim your limbs, then shall you truly dance.*

When you drink from the river of silence. Drink from the river of silence.

Lucy had. Not in the way the author inferred. But she knew what it was to plunge into a river of silence and find a new way to sing.

Another reprise of the chorus. She signed *drink, river, silence, sing*... Simple signs she'd practiced hard to master. She and Caden sang their duet, though she had no voice with which to sing.

Evan hadn't tolerated the press of the crowd when the service ended. So they exited as quickly as they'd come and had been home an hour already before Charlie and Lucy arrived.

"That was a beautiful service," Sam said. "Kind of surprisingly, I have to say. And Mom. Nice."

"Thank you."

Sasha stood from where she and Sam had been sitting, arms entwined, on the couch. She approached Lucy slowly. Sasha picked up Lucy's hands, held them in her own, then bent to kiss them, as if blessing them. She hugged Lucy tight and signed the now familiar, "I love you."

Lucy wasn't sure if she wanted to know the answer, but asked, "Did you get the apartment?"

Sam looked at Sasha, who had returned to the nest of Sam's embrace, and shook his head no.

Charlie removed his tie and said, "Something will turn up. You can't put this much prayer behind something without an answer showing up eventually." He turned away mid-sentence to kick off his shoes, so Sam signed the rest of it for his wife.

"Evan sleeping?" Lucy made sure to face Sasha when she asked.

Yes.

"Are you hungry?"

Sam rubbed his hands across his midsection. "We stopped at the drive-thru on the way home."

Charlie turned. "I could be talked into something to eat."

"We ate at the church." Lucy found his subdued enthusiasm comical.

"Then we had that long drive."

Charlie, Charlie, Charlie.

"I'm going to make some hot tea. Anybody?" Lucy glanced at her family for responses. No takers.

Charlie headed for the bedroom to change clothes.

Sam called into the kitchen, "How long before Olivia and Caden get here?"

"The interment isn't until the weekend. So they'll come straight here after the last of the guests leave. Maybe another hour at the most." Lucy cringed at the roughness that tinted everything she said. She missed the voice that calmed her children, read them stories, sang them songs. This voice wouldn't calm anyone.

"Okay."

Lucy took her tea with her to the bedroom to change. Charlie helped her with the zipper, then held her for a long moment.

"LucyMyLight, that was so worshipful."

"You couldn't have said anything more meaningful to me, Charlie."

He nestled his mouth close to her ear and sang, "You are my sunshine. My only sunshine." The whole song. His voice melted into her.

"Are you trying to sing me to sleep?"

"Later," he said.

She slipped out of her dress and into jeans and a cotton shirt while Charlie finished buttoning his plaid lightweight flannel. "I love your voice, Charlie. Why haven't you sung to me more often?"

"Oh, you were the one with the voice. We all knew that."

"Charlie. Will you sing me to sleep later? For real?"

He hugged her again. "Anything for you."

Within the hour, Caden and Olivia arrived. Caden seemed visibly relieved the service was over.

Sam and Sasha made room on the couch for the other young couple. Lucy took her reading chair. Charlie claimed his recliner. The small talk dwindled to a peaceful silence, the only noise coming from the squeaky spring in Charlie's chair.

Sam leaned forward. Sasha put her hand on his back.

"We didn't get the apartment," he said, directing his words to Caden and Olivia. "Nothing we've seen is big enough for us."

"We're so sorry," Olivia said. "I keep checking with the people I know around here with rentals."

He glanced at Sasha. "We're going to need a bigger place." He emphasized the word *bigger*.

"Well," Charlie said, "you might have to reconsider how much you're willing to pay per month then. Or do what I suggested and look for a home to buy and—"

Lucy leaned forward too. "How big?" She watched Sasha's eyes. Lucy clasped her arms in front of her and moved them left to right, right to left. "This big?"

Sasha nodded.

"You're having a baby?" Olivia jumped off the couch and wrapped her arms around both Sam and Sasha.

"What?" Charlie said. "Where'd you get that? They need a bigger apartment so I said they should think about—"

Lucy made the international sign again. "Charlie, they're expecting a baby."

Sam sighed and tightened his grip on Sasha's hand. "It's not great timing."

Lucy wondered how few pregnancy announcements started with the words, "This is great timing." Then reality settled over her. Evan had made so much progress. How would a baby in the house change the dynamic for that little man? Charlie might be needed more than ever to keep Evan as stable as possible. Those two had the kind of connection Asperger's researchers would probably appreciate studying. "We're excited for you two. You three," Lucy corrected. "And we're here for you."

"You always are. Thanks. We're focusing on the beautiful side of this unexpected news. If we don't, it gets a little overwhelming."

How different this pregnancy would be from Sasha's last, but still threaded with concerns.

Sasha leaned against her husband's shoulder.

Sam closed his eyes for a moment. "We didn't plan to drop this news on a day like today, Caden."

"Hey, man. It's the perfect ending to this day. I'm thrilled for you. Really." When Olivia came back to earth, he asked if he could get a drink of water.

The poor guy. Such highs and lows. He'd seemed to weather the funeral well. He still had the interment and all the legal things, paperwork, his mother's estate to figure out. Lucy loved watching Olivia provide him that safe place to land.

She made herself a note to add that to her "The Wife I Want to Be" list.

Caden left to get a few things done at the house. Olivia retreated to her room, but came out when the pizza delivery guy showed up at suppertime.

Evan's slice of pizza had been stripped of everything—sausage, cheese, sauce. But he ate it. Sam and Sasha leaned on each other through the whole meal. Lucy knew the sign for *happy* and used it often as they ate.

Lucy's swallowing had advanced to the stage where she could eat pizza, if she cut it into tiny bits and chewed with gusto—*gustoso*. She drank lots of water to aid the process. It felt good to enjoy strong flavors again.

"Are you working tomorrow, Sam?" Olivia picked at a slice of perfectly translucent onion.

"No. I took both of these days off for the apartment search and doctor's appointment."

"Caden would like to host us for breakfast at his mom's place in the morning. Can everybody come?"

Lucy had always been curious about what Mrs. Schindler's property looked like beyond its gates and the winding driveway that hid the house. She expected it to be cold and sterile, and then apologized silently for profiling again. "I'm free," she said and looked Charlie's way.

"You know me," he said. "I'm home all the time."

Lucy had a hard time not spitting her water back into her glass.

Sam said Sasha was fine with it but wondered if she should bring Evan or stay home with him. Olivia assured her Evan was welcome, too. A simple breakfast-ish thing, she said.

"If nine's okay, I won't have to call Caden back," she said.

Lucy sighed. "He's going to have to get on the road soon, isn't he?"

Olivia dropped her gaze to her lap. "This is his last road trip for a while. He has more than he can handle with his duties at the Performing Arts Center. He had to choose one or the other. I'm happy, obviously, that he'll be in the area more. But performing is in his blood."

Evan signed *happy*, then went back to denuding another slice of pizza.

"Ironic, isn't it? Working the youth music programs at the Performing Arts Center keeps him from performing as often as he'd like."

Lucy rehearsed the impact of his music at the funeral. "Gifted man."

Olivia sighed.

"I do believe my sister's in l-o-v-e," Sam said.

"I admit to having...feelings...for the gentleman," she said.

Lucy rested her elbows on the table, her chin in her hands.

"Are you okay, Lucy?"

"Listening to the sounds of life." And dreaming. Was she ready to share what the dream looked like?

40

Evelyn Schindler had lived in a style surrounded by warmth that made Lucy wonder if they had the right address. The gates were more ornamental than protective or exclusionary. The driveway curved, but wasn't long and eerie, as Lucy suspected. Her modest two-story home sat on a hill overlooking the Willow River. Lucy tried to calculate if she and Charlie might have canoed past it on their now-famed voyage.

Caden welcomed them on a day as crisp and clear as the previous day had been foggy. Temps were cool enough for jackets, the hills brilliant with autumn colors. Lucy knew she wasn't the only one who felt awkward walking the expansive yard of a woman so recently deceased, as if her personal space were now public property. She thought she saw a flicker of the same discomfort in the deceased's son.

Caden had a breakfast buffet waiting in the vintage decorated sunroom. The food looked and smelled delicious. *He cooks, too? Olivia, hang onto this one.*

Caden seemed both at ease and out of place, if that were possible. He directed the Tuttles—absent Evan and Sasha, who was feeling every bit of the truth of the pregnancy test results—to the buffet line. He indicated their places around the sunroom table—a piece of furniture much larger than Evelyn Schindler would have needed. What must it have been like for her to have breakfast alone every morning? Her husband long gone. Her son estranged by her decree. How many mornings had she sipped her coffee and wondered if she should have handled her disappointment differently?

Lucy took extra time unfolding and positioning her cloth napkin while she considered the common ground she held with a woman she'd considered her nemesis. Evelyn was gone. Lucy still had time to chart a new course.

"What are you planning to do with this old place?" Charlie asked, a bite of blueberry muffin only partially chewed. "Is this yours now?"

"Charlie!"

Caden laid aside his napkin and pushed his chair back from the table a few inches. "For all the heartache I unintentionally caused my mother, yes, it's mine. She signed the house over to me soon after she was diagnosed." He looked around the room. "I'll probably do a little remodeling, eventually."

"If you need help with that, let me know," Charlie said.

Lucy couldn't stop her thoughts from drifting to their still-unfinished basement.

"I never considered this place would be more than a landing pad when I was touring full-time. But I couldn't pass up the opportunity at the Barre. It's more work than I can handle and keep doing gigs, though." He picked at a crumb on the tablecloth. "Time to grow up, I guess."

"What if it isn't?" Lucy said.

All eyes turned to her.

The wild swirl of midnight ideas aligned themselves like obedient puzzle pieces in the daylight. "I'd like to apply for your job. Part-time."

"Lucy?" Charlie set down his muffin. It was that serious.

"Do you think we could work together, Caden? I appreciate your vision, what you've done already."

"A lot of it came from studying what you and your father did, Mrs. Tuttle."

"If we worked together, you could still be free to take the gigs you feel most strongly about. Do both, without neglecting the Performing Arts youth programs."

Olivia's eyes glistened. She gripped Caden's hand on the table.

Lucy raised her hands as if about to call for a downbeat. "We don't know that the Center would go for the idea. And maybe it doesn't interest you, Caden." Oh, for a voice that didn't sound as if every sentence were a whine or a growl.

"They know your reputation very well, Mrs. Tuttle. Your name comes up more frequently than you'd imagine."

"Maybe we could eventually start a satellite program here in Willowcrest." Lucy felt her toes edging toward a cliff-edge marked, "Single file. Watch for falling rocks."

Caden held his forehead with one hand. "That's been on my heart for a long time. I didn't know how to pull it off."

"I have some ideas. Including integrating music into therapy for autism."

Sam put his hand on his mother's back. "This is great, Mom."

"Don't get too excited. We have a board of decision makers to get through before we can make any plans. And Charlie."

"What did I do?"

"I need to know you'd be okay with this. If it works out."

Charlie leaned back. "You didn't really wonder if you'd have my support, did you?"

"No. But..." She signed one of the ASL sentences she'd been practicing.

Charlie turned to Sam. "What did she say?"

Sam smiled. "We're in this together."

"I'm proud of you, Mom." Lucy watched her daughter's warm smile blossom as they cleared the table together.

"I'm a little proud of me, too. No, that's not the right word. I'm grateful. Life's too short to waste it waiting for it to make sense. Time to make music out of the discord."

"Resolve the chord?" Caden asked.

"Something like that." Lucy took the meat platter from his hands.

"You two don't have to do the dishes." Caden turned on the faucet and squirted liquid dish soap into the sink.

"Your mom didn't have a dishwasher?" Olivia scanned the kitchen.

"No. For all her talk about progress, she held pretty tightly to a few habits from another era. If I ever get married, I may have to—" He caught Olivia's gaze and opted not to finish the sentence.

Lucy had her preferences for how that would turn out. But it wasn't her life. It was Olivia's puzzle.

"I've entertained a crazy dream for a long time." Caden talked as the sink filled. "I didn't know how it would happen or when, but I see the pieces coming together."

"What kind of dream?" Lucy's skin didn't prickle at the mention of that word anymore. Another victory.

"A music camp. Here. On this property."

"I wouldn't call that crazy," Olivia said.

Lucy lowered a handful of silverware into the sudsy water and stared out the window into the backyard that sloped gently toward the river. "Can you imagine how the sound would carry here?"

Lucy curled against Charlie, her head on his chest. She listened to his heartbeat, that magnificent rhythm God invented. Efficient, steady. Yet expressive of emotion and tension at times. Tonight it beat slowly as he ran his fingers through her hair. She needed a haircut. So did he. Maybe they'd go together.

"Charlie?"

"Um hmm."

"Are you excited about Sam and Sasha's new baby?"

"Sure. Aren't you?"

"Thrilled for them. And concerned for them, too. Space is only one of the issues they'll face. Evan..."

"They'll make it."

He said it with the confidence of a man who had seen miracles.

Lucy twisted to look at his face. "Do you know what I've been thinking about?"

"What, Lucy?"

"Zechariah, the prophet."

"That's what I was going to guess," Charlie said.

She poked his belly. "I'm trying to be serious and all spiritual here."

"Sorry. Go on."

"John the Baptist's dad."

"I remember."

"When the angel told him that his wife, Elizabeth, was going to have a baby, Zechariah didn't believe God could do such a thing."

"Um hmm."

"Are you sleeping?"

"Nope. Not at all."

Lucy snuggled in tighter. "He wasn't able to speak the whole time Elizabeth was pregnant, because he hadn't believed. He wasn't able to *speak*, Charlie."

"Yup."

"He had to write what he wanted to say on tablets. Like a biblical version of a whiteboard."

Charlie yawned. "Sounds familiar."

Lucy matched his yawn. "When he wrote on the board that the baby's name was John, because of what God had told him through the angel, he instantly got his voice back."

"Yours didn't come back instantly."

"Stick with me here, Charlie. I'm making a point."

"Okay."

"The verse in Luke that describes the scene says that as soon as he could speak again, he began praising God."

Charlie cleared his throat. "Like you."

"That's not exactly the point I was trying to make either. I think I've had kind of a Zechariah summer. Charlie?"

"Listening. Completely. Full attention."

"What would you think if I started writing?"

"What do you mean? Like, books?"

Her sigh moved them both. "I've been thinking about writing a book for first-year music instructors. A lot of practical things for survival, hints on classroom discipline, how to break through a student's resistance, organizational tips..."

He shifted so he could tilt her face toward his. "I love the idea."

"I've taken notes for years. Every student teacher who's worked with me has left with a notebook full of thoughts to make their first year on the job as meaningful and trouble-free as possible."

"Sold."

"You think it would work? Do you think I could pull it off?"

"No doubt in my mind."

"And then," she said, "my second book would be about musical literacy, including converting schoolwork into song as a learning aid, like with Evan. And then the third could be used by music programs that are threatened with extinction to present a comprehensive look at the value of music in creating well-rounded students who excel."

"I'll buy a copy."

Lucy thought about reaching for her earbuds. "Charlie?"

"Yes, LucyMyLight?"

"Will you sing me to sleep?"

Group Discussion Guide

1. One of the themes of *Song of Silence* dealt with the loss of Lucy's sense of identity when she lost her teaching position, or rather, her passion. What in the way she responded was unhealthy? Was there anything healthy in her reaction?

2. How would it have changed things if Lucy had sought professional help sooner? Or did she have to be ready to receive the help offered first? What does that say to us when we're reaching out to people in pain?

3. Charlie's and Lucy's expectations for what retirement was supposed to look like were in some ways at opposite ends of the spectrum. Some have suggested that counseling couples on the verge of their retirement years is as important as premarital counseling. How might that have benefited Lucy and Charlie? Where did their expectations collide first in the story?

4. Rests—pauses—play key roles in *Song of Silence*. Some are welcomed. Others feel like cruel interruptions. What important pauses do you recognize in the story? How have you seen "timed rests" playing an important part in the lives of those around you?

5. If music is a vital part of your life, you probably identified strongly with the way it's used in the story to express longing, joy, despair, heartache, hope, healing. How did music's role—a character itself—change from the beginning of the story to the end?

6. Some disappointments can be shrugged off. Others pierce us to the core. Each of the characters in *Song of Silence* took a unique path in dealing with his or her disappointments. With which character's responses did you most closely identify?

7. One of Lucy's deepest concerns was to know that her life's work mattered, that she'd made a difference with her efforts. Do you believe that's a universal longing? Or is that drive stronger in some than others? How did that drive show itself in Charlie, for instance?

8. Charlie and Lucy faced their grown children needing to come home again, temporarily. What effect did that have on the story's turning points? If you've had a similar experience with your grown children (or if you've been the one who needed to move home for a while), how did you make it work?

9. The thing that causes the greatest happiness can also be the source of the greatest pain. That played out in more than one arena in *Song of Silence*. Which ones stood out to you?

10. Describe the differences between the friendship Charlie had with Martin and the one Lucy developed with her book club friends. Why were Lucy's friends successful in helping her get her feet underneath her again?

11. Which tender scene in the story resonated most strongly with you? Why?

12. Lucy and Charlie have been married many years when the story opens. What surprising things did Lucy learn about Charlie before the final pages? What difference did those revelations make for their future together?

A NOTE FROM THE EDITORS

We hope you enjoyed *Song of Silence* by Cynthia Ruchti, published by the Books and Inspirational Media Division of Guideposts, a nonprofit organization that touches millions of lives every day through products and services that inspire, encourage, help you grow in your faith, and celebrate God's love.

Thank you for making a difference with your purchase of this book. Your purchase helps support our many outreach programs to military personnel, prisons, hospitals, nursing homes, and educational institutions.

We also create many useful and uplifting online resources. Visit Guideposts.org to read true stories of hope and inspiration, access OurPrayer network, sign up for free newsletters, download free e-books, join our Facebook community, and follow our stimulating blogs.

To learn about other Guideposts publications, including the best-selling devotional Daily Guideposts, go to Guideposts.org/Shop, call (800) 932-2145, or write to Guideposts, PO Box 5815, Harlan, Iowa 51593.

Sign up for the Guideposts Fiction Newsletter

and stay up to date on the books you love!

You'll get sneak peeks of new releases, recommendations from other Guideposts readers, and special offers just for you . . .

and it's FREE!

Just go to Guideposts.org/Newsletters today to sign up.

Guideposts.

Visit Guideposts.org/Shop
or call (800) 932-2145

Find more inspiring fiction in these best-loved Guideposts series!

Tearoom Mysteries Series
Mix one stately Victorian home, a charming lakeside town in Maine, and two adventurous cousins with a passion for tea and hospitality. Add a large scoop of intriguing mystery and sprinkle generously with faith, family, and friends, and you have the recipe for *Tearoom Mysteries.*

Sugarcreek Amish Mysteries
Be intrigued by the suspense and joyful "aha" moments in these delightful stories. Each book in the series brings together two women of vastly different backgrounds and traditions, who realize there's much more to the "simple life" than meets the eye.

Mysteries of Martha's Vineyard
What does Priscilla Latham Grant, a Kansas farm girl know about hidden treasure and rising tides, maritime history and local isle lore? Not much—but to save her lighthouse and family reputation, she better learn quickly!

Mysteries of Silver Peak
Escape to the historic mining town of Silver Peak, Colorado, and discover how one woman's love of antiques helps her solve mysteries buried deep in the town's checkered past.

To learn more about these books, visit Guideposts.org/Shop